Nečas Center Series

T0181002

Birkhäuser

The Nečas Center Series aims to publish high-quality monographs, textbooks, lecture notes, habilitation and Ph.D. theses in the field of mathematics and related areas in the natural and social sciences and engineering. There is no restriction regarding the topic, although we expect that the main fields will include continuum thermodynamics, solid and fluid mechanics, mixture theory, partial differential equations, numerical mathematics, matrix computations, scientific computing and applications. Emphasis will be placed on viewpoints that bridge disciplines and on connections between apparently different fields. Potential contributors to the series are encouraged to contact the editor-in-chief and the manager of the series.

More information about this series at http://www.springer.com/series/16005

Roman Shvydkoy

Dynamics and Analysis of Alignment Models of Collective Behavior

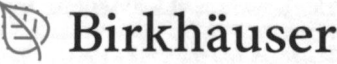 Birkhäuser

Roman Shvydkoy
Department of Mathematics,
Statistics, & Computer Science
University of Illinois at Chicago
Chicago, IL, USA

ISSN 2523-3343   ISSN 2523-3351   (electronic)
Nečas Center Series
ISBN 978-3-030-68146-3   ISBN 978-3-030-68147-0   (eBook)
https://doi.org/10.1007/978-3-030-68147-0

Mathematics Subject Classification: 92D15, 35Q35, 76N10

This book is published under the imprint Birkhäuser, www.birkhauser-science.com by the registered
company Springer Nature Switzerland AG
The registered company address is: Gewerbestrasse 11, 6330 Cham, Switzerland

To Olga, Dmitriy, and Uliana

Preface

Collective phenomena occur in nature, technology, and social behavior in a vast variety of contexts. For example, birds and fish use their sensory abilities to communicate and align their positions to form cohesive congregations—flocks and schools, respectively. Migration of microorganisms such as bacteria or cells follows primitive bio-chemical communication rules to induce collective motion that is crucial to sustain life in more complex organisms. In technology, communication between agents to achieve common collective goals is the problem that spans across many different applications such as control of unmanned aerial vehicles, coordinated satellite navigation, traffic control, etc. Social science offers another variety of examples of collective behavior such as dynamics of opinions and reaching consensus, social and economic networks, and emergence of leaders. With the advent of new techniques and methods in PDEs, analysis of mathematical models of collective motion is starting to become one of the most actively developing subjects of applied science. This book is a modest attempt to introduce the reader to one special class of alignment models, called Cucker–Smale system, and its kinetic and hydrodynamic counterparts, which has undergone tremendous development in recent years. From a purely analytical prospective, one of the most fascinating aspects of these systems is that they bring together a toolbox of several modern techniques that emerged recently in fluid dynamics, fractional analysis, and kinetic theory. Our primary goal will be to walk through all levels of description—discrete, kinetic, and hydrodynamic—in a mostly self-contained manner, and to highlight essential analytical aspects at each step. Those include regularity theory and long time behavior of solutions. For the most part, we will be concerned with a "bare bone" model, where the main structure is stripped of other accessories such as external forces, various modifications, etc., except for a special class of classical potential and friction forces due to their close relevance to applications.

The material of this book represents a modified and largely streamlined content of several works, many of which are very recent and are still in production. All relevant references and discussions will be provided at the end of each chapter. We will by no means attempt to provide a comprehensive coverage of the selected topics, but the reader will find these surveys very useful for further guidance on the subject [1, 14, 17, 73, 76, 102].

Our starting point is the agent-based Cucker–Smale system that we introduce in Chap. 1. We will briefly discuss the historical context and a range of relevant applications. A rough classification of collective phenomena and most frequently occurring types of communication will be given. The basic question here is how much communication is necessary to induce unconditional alignment? We present two methods of approach to long time behavior—the original Cucker–Smale spectral method and the Lyapunov function approach of Ha and Liu. These methods will reemerge in both kinetic and hydrodynamic contexts later in the text. A first major deviation from the classical non-degenerate kernels will be provided with the introduction of the corrector method, which allows us to include more realistic communication protocols, for example, as a part of multi-zone models to be discussed in Chap. 3. We give a brief overview of existing multi-scale models at the end of Chap. 2.

Although in most of the text we focus on the core alignment mechanisms omitting any forces, in Chap. 3 we provide a discussion of a few forced models because of their relevance to applications and recent revelations in understanding their role in alignment dynamics. The central case studied here is the potential repulsion–attraction forcing and the multi-zone analysis that comes with it. As in the degenerate kernel case, the energy law in the presence of an interaction force suffers from a lack of coercivity. We present a new adaptation of the hypocoercivity method to tackle this issue. Dynamics under Rayleigh friction and self-propulsion in the context of Cucker–Smale alignment will be discussed in Sect. 3.4.

The analysis of kinetic alignment models—a Vlasov-type direct counterpart of Cucker–Smale system—will be the subject of Chap. 4. The kinetic formulation will be derived formally via the BBGKY hierarchy and justified rigorously through the mean-field limit. The analysis of the limit comes with additional contractivity estimates that in the case of heavy tail communication automatically show the stability of flocking dynamics. The passage from kinetic to hydrodynamic description will be discussed in Chap. 5. Our approach is based upon a kinetic version of the relative entropy method implemented for flocking models by Kang, Figalli, and Vasseur with a few modifications. In particular, we rely on a smooth kinetic model that can be justified via a mean-field limit, and we formulate the results for compactly supported flocks in the open space.

The hydrodynamic version of the Cucker–Smale system, called Euler alignment system, will be discussed in detail in Chap. 6. In its Lagrangian description, the system has a very similar structure to its agent-based counterpart. As a result, a number of statements carry over from the microscopic to the macroscopic description verbatim. We will introduce a class of topological models to address the deficiencies of the classical metric models in the case when communication is local. Here, the hydrodynamic connectivity interacts with the adaptive diffusion built into a topological protocol to produce natural flocking results. Subsequent analysis of such models will not be included because of technical complexities that go beyond the scope of this text.

In Chaps. 7 and 8, we provide the core well-posedness theory of both smooth and singular models. We start with local existence results supplemented by proper continuation criteria, which will be useful to develop global regularity theory for one-dimensional (1D) and some multi-dimensional systems. Understandably, the analysis of smooth models will be quite different from the analysis of singular models because of the differences in the type of PDEs we are dealing with. Smooth models exhibit the structure of a hyperbolic system of conservation laws, for which one can provide threshold criteria for regularity in terms of an entropy-like quantity. Singular models fall into the class of fractional parabolic equations with rough drift, which has undergone major development in recent years in the context of fluid dynamics (critical SQG, fractional Burgers, etc.). We will present a streamlined proof of the main regularity result for 1D models introduced by Tadmor and the author in the trilogy of works [93–95]. Here, we recall relevant tools of analysis of critical fractional PDEs, making the exposition mostly self-contained with the exception of Nash–Moser-type regularization result of Silvestre [98] that would carry us beyond the scope of the book. For the flocking analysis of singular models, the key technical ingredient will be an adaptation of the non-local maximum principle of Constantin and Vicol [28]. An independent approach to regularity via the use of modulus of continuity method appeared also in Do et al. [39]. Although it is a powerful tool for studying fractional models, we will not cover it for the sake of brevity and refer to [2, 39, 62] for its appearance in the context of flocking.

A few partial results are known concerning the regularity of multi-dimensional Euler alignment systems; however, the theoretical basis in this case remains largely underdeveloped. We present them in Chap. 9. Those include small initial data for smooth models in terms of the spectral gap of the initial condition, and for singular models in terms of the amplitude of velocity oscillations. We describe a new class of unidirectional solutions and prove their regularity and stability.

Several pressing open problems will be highlighted throughout the text and in the Notes and References sections.

The material of this book is based upon several series of lectures given at the ICMAT in November 2018, Charles University in May 2019, and Beijing Normal University in July 2019, as well as a working group seminar at UIC. The author is grateful to Josef Málek for encouraging him to write and publish these notes in the Nečas Center Series. Constant support of NSF, Simons Foundation, and College of LAS at UIC during preparation of the manuscript is gratefully acknowledged. Special thanks go to Eitan Tadmor for introducing the author to the subject with unmatchable openness and limitless enthusiasm.

Chicago, IL, USA   Roman Shvydkoy
October, 2020

Acknowledgments

Author's research is supported in part by NSF grant DMS-1813351 and Simons Foundation.

Contents

Chapter 1
Emergent Phenomena and Overview of Existing Models

Generally speaking, *emergence* is a phenomenon of reaching collective outcome in a given system of agents which follow a prescribed protocol of communication. Some of the most striking examples of emergence are pattern formations such as wedges of bird flocks, milling motion in schools of fish, and lattices in cell organization or beehives. These effects are achieved by following local rules of engagement that result in global outcomes over time. Depending on the context, many mathematical models have been studied to replicate a specific collective behavior. Roughly, most classes of models fall into two categories - first-order systems and second-order systems. First-order systems often model evolution of non-inertial flocks, those that do not induce motion if no force is present. For example, in studying exchange of opinions or networking, one of the most popular models used is environmental averaging:

$$\dot{\mathbf{p}}_i = \lambda \sum_{j \in \mathcal{N}_i} a_{ij}(t)(\mathbf{p}_j - \mathbf{p}_i) + \mathbf{F}_i, \qquad \sum_j a_{ij}(t) = 1,$$

where N_i is a set of "active" agents in local proximity and $\mathbf{p}_i \in \mathbb{R}^n$ stands for a characteristic state of i'th agent such as its opinion. The main alignment term represents the statistical averaging over actively involved agents, and $\mathbf{F}_i$ is an external random or deterministic force to account for possible additional effects. The typical emergent phenomenon here is achieving consensus, $\mathbf{p}_i \to \bar{\mathbf{p}}$; see [37, 76, 79] for further reading.

A large class of first-order gradient models appears in numerous biological and physical applications such as particle dynamics and cell migration and is given by

$$\dot{\mathbf{x}}_i + \sum_{j:j \neq i} \nabla_{\mathbf{x}_i} W(\mathbf{x}_i - \mathbf{x}_j) = 0, \quad i = 1, \ldots, N, \tag{1.1}$$

© Springer Nature Switzerland AG 2021
R. Shvydkoy, *Dynamics and Analysis of Alignment Models of Collective Behavior*,
Nečas Center Series, https://doi.org/10.1007/978-3-030-68147-0_1

where W is a radially symmetric repulsion/attraction potential and $\mathbf{x}_i$'s are spatial positions of the agents typically on a given closed surface $\Sigma \subset \mathbb{R}^n$ or simply in $\mathbb{R}^n$. The most well-studied collective outcomes are lattice patterns in the distribution of global minimizers of potential energy.

One famous example of a gradient system is given by Kuramoto's synchronization model:

$$\dot{\theta}_i = \frac{\lambda}{N} \sum_{j \in \mathcal{N}_i} \sin(\theta_j - \theta_i) + \omega_i, \quad \theta_i \in \mathbb{T}^1,$$

where θ_i are phase angles of agents and ω_i are prescribed natural frequencies. It appears in a surprisingly diverse array of examples, such as neuronal signals in the brain, simulating cardiac pacemaker cells, synchronization of power grids, and even cricket pitches in the garden; see [1, 64]. The model exhibits phenomenally complex behavior despite its relative simplicity. The size of coupling strength relative to natural frequencies may trigger phase transitions from chaotic to synchronous.

A class of models that take into account inertial effects is models of second order describing evolution of pairs of phase points:

$$\mathbf{x}_i \in \Omega \subset \mathbb{R}^n, \quad i = 1, \dots, N$$

$$\mathbf{v}_i = \dot{\mathbf{x}}_i.$$

One of the most studied models in biological literature is the time-discrete Vicsek model:

$$\left\{ \begin{aligned} &\mathbf{v}_i(k+1) = v_0 \frac{\sum_{j:|x_j - x_i| < r_0} \mathbf{v}_j}{\left| \sum_{j:|x_j - x_i| < r_0} \mathbf{v}_j \right|} + \mathbf{F}_i, \\ &\mathbf{x}_i(k+1) = \mathbf{x}_i(k) + \mathbf{v}_i(k+1). \end{aligned} \right.$$

Depending on the nature of the forces $\mathbf{F}_i$, solutions may form interesting flocking patterns such as mills or periodically rotating chains with or without self-intersections. The model undergoes phase transitions from disordered to ordered state, depending on the level of noise present in the system; see [102] for comprehensive discussion.

In this book, the center of our attention will be the analysis of alignment systems, the prototypical example of which is the Cucker-Smale second-order model given by

$$\left\{ \begin{aligned} &\dot{\mathbf{x}}_i = \mathbf{v}_i, \\ &\dot{\mathbf{v}}_i = \frac{\lambda}{N} \sum_{j=1}^N \phi(\mathbf{x}_i - \mathbf{x}_j)(\mathbf{v}_j - \mathbf{v}_i), \end{aligned} \right. \qquad (\mathbf{x}_i, \mathbf{v}_i) \in \Omega \times \mathbb{R}^n, \qquad (1.2)$$

where $\phi(\mathbf{x}) = \phi(|\mathbf{x}|)$ is a communication kernel, or influence function, originally chosen to be

$$\phi(r) = \frac{1}{(1 + r^2)^{\beta/2}}, \quad \beta > 0. \tag{1.3}$$

The system was introduced in [33, 34] in response to the need for a model which does not require perpetuating assumptions of connectivity of the flock to ensure unconditional alignment, a deficiency of local models. The model has seen instant success, thanks to its simplicity and amenability to analysis. A Cucker-Smale protocol with calibrated value of $\beta = 0.4$ was proposed to be used in satellite communication in the Darwin mission [81]. Alignment dynamics is amenable to control problems and sustains even degenerate communications [12, 38]. Prescribed collective outcomes can be achieved via decentralized control [9, 26]. Applications were found to meta-heuristic optimization algorithms [18]. The flexibility of the Cucker-Smale model allows one to incorporate individual characteristics of agents through thermodynamic parameters [44]. Flocking behavior appears relevant in modeling hybrid agent systems embedded in an incompressible fluid [47] and in multi-scale and multispecies systems [53, 65, 96]. Other features of Cucker-Smale dynamics based on hierarchy, angle of vision, and emergence of leaders are reviewed in [14].

Chapter 2
Agent-Based Alignment Systems

In this chapter, we study a general class of agent-based systems of Cucker-Smale type:

$$
\begin{cases}
\dot{\mathbf{x}}_i = \mathbf{v}_i, \\
\dot{\mathbf{v}}_i = \lambda \sum_{j=1}^{N} m_j \phi(\mathbf{x}_i, \mathbf{x}_j)(\mathbf{v}_j - \mathbf{v}_i),
\end{cases}
\qquad (\mathbf{x}_i, \mathbf{v}_i) \in \Omega \times \mathbb{R}^n. \qquad (2.1)
$$

Here, Ω denotes the environment, and m_j are "masses" of agents. We always consider either open space $\Omega = \mathbb{R}^n$ or a periodic domain $\mathbb{T}^n$. The meaning of mass, however, depends on the context. It could be the actual physical mass of an agent, or it could be its strength to influence others. For most of our exposition, we omit any forces, focusing more on the core dynamics of self-organization.

This chapter lays out a framework for a subsequent discussion introducing basic terminology and classification of communication kernels and presenting the most basic alignment results. We also introduce the corrector method which allows us to expand the analysis to degenerate communications, to be applied later to forced models.

2.1 Types of Communication and Collective Outcomes

For a given solution $\{\mathbf{x}_i, \mathbf{v}_i\}_{i=1}^{N}$ to system (2.1), we identify the following collective outcomes of long time behavior (see Fig. 2.1):

© Springer Nature Switzerland AG 2021
R. Shvydkoy, *Dynamics and Analysis of Alignment Models of Collective Behavior*,
Nečas Center Series, https://doi.org/10.1007/978-3-030-68147-0_2

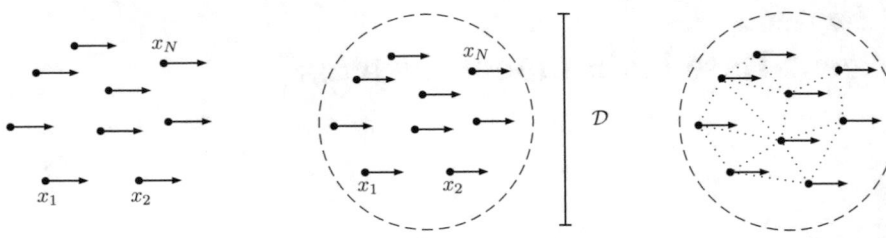

Fig. 2.1 Alignment, flocking, and strong flocking

- Alignment: $\lim\limits_{t\to\infty} \max\limits_{i} |\mathbf{v}_i - \bar{\mathbf{v}}| = 0,$

- Flocking: $\sup\limits_{i,j} |\mathbf{x}_i - \mathbf{x}_j| \leq \overline{\mathcal{D}} < \infty,$

- Strong flocking: $\mathbf{x}_i - \mathbf{x}_j \to \bar{\mathbf{x}}_{ij}, \text{ as } t \to \infty,$

- Aggregation: $\mathbf{x}_i - \mathbf{x}_j \to 0, \text{ as } t \to \infty.$

In fact, aggregation is not typical for forceless dynamics and will become relevant later in Chap. 3 when we consider attraction forces.

Note that alignment implies strong flocking provided it occurs at a sufficiently fast rate. Indeed, if

$$\int_0^\infty \max_{i,j} |\mathbf{v}_i - \mathbf{v}_j| \, dt < \infty, \tag{2.2}$$

then

$$\mathbf{x}_i(t) - \mathbf{x}_j(t) = \mathbf{x}_i(0) - \mathbf{x}_j(0) + \int_0^t [\mathbf{v}_i(s) - \mathbf{v}_j(s)] \, ds;$$

hence,

$$\bar{\mathbf{x}}_{ij} = \mathbf{x}_i(0) - \mathbf{x}_j(0) + \int_0^\infty [\mathbf{v}_i(s) - \mathbf{v}_j(s)] \, ds.$$

This is why it is important to provide a rate of alignment whenever possible.

Properties of the communication kernel ϕ play a crucial role in alignment dynamics and may dramatically change the collective behavior of the system. We always assume that the kernels are of alignment type, i.e.,

$$\phi(\mathbf{x}, \mathbf{y}) \geq 0, \qquad \forall \mathbf{x}, \mathbf{y} \in \Omega.$$

We distinguish between the following general types of communication:

- Absolute: $\inf\limits_{\mathbf{x},\mathbf{y}\in\Omega}\phi(\mathbf{x},\mathbf{y}) > 0,$

- Non-degenerate: $\phi(\mathbf{x},\mathbf{y}) > 0,\ \forall\mathbf{x},\mathbf{y}\in\Omega,$

- Local: $\exists\, r_0 > 0$ so that $\phi(\mathbf{x},\mathbf{y}) = 0$ for $|\mathbf{x}-\mathbf{y}| > r_0,$

- Symmetric: $\phi(\mathbf{x},\mathbf{y}) = \phi(\mathbf{y},\mathbf{x}),$

- Convolution type: $\phi(\mathbf{x},\mathbf{y}) = \phi(\mathbf{x}-\mathbf{y}),$

- Radial: $\phi(\mathbf{x},\mathbf{y}) = \phi(|\mathbf{x}-\mathbf{y}|).$

Sometimes, by a local kernel, we simply mean the lack of global assumptions, for example, when we state

$$\phi(\mathbf{x},\mathbf{y}) \geq 1,\ \text{ for } |\mathbf{x}-\mathbf{y}| \leq r_0. \tag{2.3}$$

The majority of our examples will be of radial convolution type in which case we also assume that $\phi(r)$ is sufficiently smooth for $r > 0$. We say that ϕ is *smooth*, and by extension, the model (2.1) is smooth if $\phi \in C^2(\mathbb{R}^n)$, i.e., ϕ is regular at the origin also. Otherwise, we say that the kernel and the model are singular. The original example used by Cucker and Smale in their seminal works is of smooth convolution type:

$$\phi(r) = \frac{1}{\langle r\rangle^{\beta}}, \qquad \langle r\rangle = (1+r^2)^{\frac{1}{2}}. \tag{2.4}$$

The most important example of a singular kernel for us will be the power kernel :

$$\phi(r) = \frac{h(r)}{r^{\beta}}, \tag{2.5}$$

where h is a possible cutoff function if we consider local kernels. If $\beta = n + \alpha$, $0 < \alpha < 2$, then this becomes the kernel of a localized fractional Laplacian, which we will study in detail in Chaps. 7 and 8. The strength of communication, with or without any presumed structure of the kernel, can be expressed in terms of an integrability condition either at the close range or long range:

$$\text{Long range heavy tail:} \quad \int_{r_0}^{\infty} \phi(r)\,dr = \infty, \tag{2.6}$$

$$\text{Short range heavy tail:} \quad \int_{0}^{r_0} \phi(r)\,dr = \infty. \tag{2.7}$$

Two special examples of non-convolution kernels will play prominent roles in our discussion. The first is a nonsymmetric communication protocol introduced by Motsch and Tadmor [75]:

$$\psi(\mathbf{x}_i, \mathbf{x}_j) = \frac{\phi(|\mathbf{x}_i - \mathbf{x}_j|)}{\sum_k m_k \phi(|\mathbf{x}_i - \mathbf{x}_k|)}. \tag{2.8}$$

This averaging, $\sum_j m_j \psi(\mathbf{x}_i, \mathbf{x}_j) = 1$, allows to avoid deficiencies of uniform CS averaging associated with far-from-equilibrium flock configurations; see Sect. 2.8. Both singular and Motsch-Tadmor kernels are meant to emphasize local interactions over global ones whenever such communication is deemed more realistic.

So far, all presented examples belong to a class of so-called *metric* kernels, meaning that communication depends on the Euclidean distance between the agents. In some biological systems, such as flocks of starlings, communication follows a somewhat different protocol where the strength of interaction depends on the density of crowd between communicating agents. When distances are measured in terms of mass, the kernel and model are called *topological*. The example that will be discussed in detail in Sect. 6.4 is given by

$$\phi_{ij}(\mathbf{x}) = \frac{1}{d_{ij}^\tau} \psi(|\mathbf{x}_i - \mathbf{x}_j|), \qquad d_{ij} = \left[\sum_{k: \mathbf{x}_k \in \Omega_{ij}} m_k \right]^{\frac{1}{n}}, \tag{2.9}$$

where ψ is a metric component and d_{ij}^n would constitute the mass of a given communication domain Ω_{ij}. If the domain is symmetric $\Omega_{ij} = \Omega_{ji}$, then so is the kernel.

2.2   Momentum, Energy, and Maximum Principle

System (2.1) with symmetric kernels, as opposed to nonsymmetric kernels, conserves the total momentum :

$$\bar{\mathbf{v}} = \frac{1}{M} \sum_i m_i \mathbf{v}_i, \quad M = \sum_{i=1}^{N} m_i, \quad \frac{\mathrm{d}}{\mathrm{d}t} \bar{\mathbf{v}} = 0. \tag{2.10}$$

Because of conservation of momentum, the center of mass moves with a constant velocity:

$$\bar{\mathbf{x}} = \frac{1}{M} \sum_i m_i \mathbf{x}_i, \quad \frac{\mathrm{d}}{\mathrm{d}t} \bar{\mathbf{x}} = \bar{\mathbf{v}}. \tag{2.11}$$

For convolution-type kernels, conservation of momentum can be used to shift the reference frame centered at $\bar{\mathbf{x}}$ due to *Galilean invariance* of the system:

$$\mathbf{x}_i \rightarrow \mathbf{x}_i - t\bar{\mathbf{v}}, \quad \mathbf{v}_i \rightarrow \mathbf{v}_i - \bar{\mathbf{v}}. \tag{2.12}$$

So in this case, we can assume without loss of generality that $\bar{\mathbf{v}} = 0$. In general, such translation invariance is not available. Nonetheless, if alignment occurs, then necessarily all $\mathbf{v}_i \to \bar{\mathbf{v}}$. In other words, we can determine the limiting velocity from initial conditions.

All systems with radial convolution communication as well as topological and Motsch-Tadmor systems are invariant with respect to *mirror symmetries*: if we denote the coordinates of the phase parameters by

$$\mathbf{x}_i = \langle x_i^1, \ldots, x_i^n \rangle, \qquad \mathbf{v}_i = \langle v_i^1, \ldots, v_i^n \rangle,$$

then for each fixed $k = 1, \ldots, n$, the transformation

$$v_i^k \to -v_i^k, \qquad x_i^k \to -x_i^k, \quad \forall i = 1, \ldots, N \qquad (2.13)$$

produces another solution. Such systems also have the *rotational symmetries*: for any orthogonal transformation $U : \mathbb{R}^n \to \mathbb{R}^n$,

$$\mathbf{v}_i \to U\mathbf{v}_i, \qquad \mathbf{x}_i \to U\mathbf{x}_i.$$

Let us consider the following variation and dissipation functions:

$$\mathcal{V}_2 = \frac{1}{2} \sum_{i,j} m_i m_j |\mathbf{v}_i - \mathbf{v}_j|^2,$$

$$\mathcal{I}_2 = \frac{1}{2} \sum_{i,j} m_i m_j \phi(\mathbf{x}_i, \mathbf{x}_j) |\mathbf{v}_i - \mathbf{v}_j|^2.$$

The system has the classical kinetic energy as well defined by

$$\mathcal{E} = \frac{1}{2} \sum_{i=1}^N m_i |\mathbf{v}_i|^2.$$

If $\bar{\mathbf{v}} = \mathbf{0}$, then $\mathcal{V}_2 = 2M\mathcal{E}$; however, it is not prudent to use energy as a measure of alignment $\mathcal{E}$ in the nonsymmetric case, simply because we do not know if $\mathbf{0}$ would remain the momentum of the system for all time.

The following energy law is easily verified:

$$\frac{d}{dt}\mathcal{V}_2 = -2\lambda M \mathcal{I}_2. \qquad (2.14)$$

At this point, one can obtain an ℓ^2-norm-based alignment result assuming absolute communication: if $\inf \phi = c_0 > 0$, then $\mathcal{I}_2 \geq c_0 \mathcal{V}_2$, and hence,

$$\dot{\mathcal{V}}_2 \leq -2c_0 \lambda M \mathcal{V}_2.$$

Hence,

$$\mathcal{V}_2(t) \le \mathcal{V}_2(0)e^{-2c_0\lambda M t}.$$

This result provides exponential alignment "on average", specifically ℓ^2-average, which does not translate well into individual information on agents. Indeed, one only obtains

$$|\mathbf{v}_i - \mathbf{v}_j| \le \frac{1}{m_i m_j} \mathcal{V}_2(0)e^{-\delta t}. \tag{2.15}$$

This estimate clearly deteriorates in the large crowd limit $N \to \infty$ when all masses vanish $m_i \to 0$. In order to improve upon (2.15), we must resort to an ℓ^∞-norm-based argument and use the maximum principle to be discussed later in Sect. 2.4.

2.3   Connectivity and Spectral Method

In the last section, we saw that flocking behavior holds trivially under the global communication condition. At the same time if communication is local, it is easy to produce an example of two agents or disconnected flocks separated by a distance longer than their communication range that would scatter in opposite directions and not align; see Fig. 2.2. So it is clear that ultimately connectivity is the key to flocking behavior. Let us make this more precise. We say that the flock is r_0-connected if for any pair of agents $\mathbf{x}_i$ and $\mathbf{x}_j$ there exists a chain of agents $\{\mathbf{x}_{k_p}\}_{p=1}^{P_{ij}}$ with endpoints at $\mathbf{x}_i$ and $\mathbf{x}_j$ and such that all $|\mathbf{x}_{k_p} - \mathbf{x}_{k_{p+1}}| < r_0$, for all $p = 1, \dots, P_{ij}$. If the flock is r_0-connected within a communication range, $\lambda\phi(r_0) = \varepsilon > 0$, then alignment can be recovered as follows. Note that the shortest chain connecting any pair of agents has no repeated agents in the chain. Hence, every chain is limited to length N. We can then estimate $\mathcal{V}_2$, assuming for simplicity that all agents have the same mass $m_i = \frac{1}{N}$:

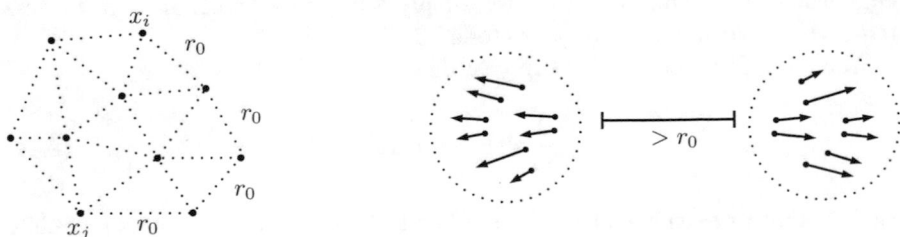

Fig. 2.2 Connected and disconnected flocks

$$\mathcal{V}_2 = \frac{1}{2N^2} \sum_{i \neq j} |\mathbf{v}_i - \mathbf{v}_j|^2 \leq \frac{1}{2N^2} \sum_{i \neq j} \sum_{p=1}^{P_{ij}} |\mathbf{v}_{k_p} - \mathbf{v}_{k_{p+1}}|^2$$

$$\leq \frac{\lambda}{2\varepsilon N^2} \sum_{i \neq j} \sum_{p=1}^{P_{ij}} \phi(\mathbf{x}_{k_p} - \mathbf{x}_{k_{p+1}}) |\mathbf{v}_{k_p} - \mathbf{v}_{k_{p+1}}|^2$$

$$\leq \frac{N(N-1)}{2\varepsilon N^2} \lambda \sum_{k' \neq k''} \phi(\mathbf{x}_{k'} - \mathbf{x}_{k''}) |\mathbf{v}_{k'} - \mathbf{v}_{k''}|^2 \lesssim N^2 \mathcal{I}_2.$$

The desired differential inequality follows:

$$\dot{\mathcal{V}}_2 \leq -\frac{1}{N^2} \mathcal{V}_2.$$

The argument produces a bad dependence on N and as such is not suitable in the limit $N \to \infty$. A continuous analogue of the connectivity condition requires more elaboration and is discussed in [74].

Connectivity at all times is hard to verify of course. However, it is guaranteed to hold provided the initial configuration is connected and the communication strength λ is large enough. This is because one can ensure in this case that the system aligns almost instantaneously before any disconnection becomes possible. Indeed, suppose initially the system is $r_0/2$-connected and $\lambda\phi(r_0)$ is large. Then for some large $\Lambda > 0$, we will have

$$\frac{\mathrm{d}}{\mathrm{d}t} \mathcal{V}_2 \leq -\Lambda \mathcal{V}_2,$$

for as long as the system is r_0-connected. So if a pair $\mathbf{x}_i, \mathbf{x}_j$ is initially at most $r_0/2$ apart, then

$$|\mathbf{x}_i(t) - \mathbf{x}_j(t)| \leq |\mathbf{x}_i(0) - \mathbf{x}_j(0)| + \frac{C}{\Lambda}.$$

This shows that the same pair will never get r_0-disconnected, and hence, the connectivity is preserved at all times.

If the kernel has a long range tail, $r_0 = \infty$, then obviously every flock is r_0-connected. But how strong that long range communication needs to be in order to actually ensure flocking? Let us illustrate the answer by the following example.

Example 2.1 Let the kernel be $\phi(r) = \frac{1}{r^\beta}$ for $r > r_0$, for simplicity, and let $x = x_1 = -x_2 > r_0$ and $v = v_1 = -v_2 > 0$. This symmetry is preserved in time. Then system (2.1) becomes

$$\frac{\mathrm{d}x}{2^{\beta-1}x^\beta} + \frac{\mathrm{d}v}{1} = 0.$$

This equation admits a conservation law:

$$J = v + \frac{1}{2^{\beta-1}(1-\beta)x^{\beta-1}}.$$

If $\beta > 1$ and the initial velocity is large enough, then $J(t) = J_0 > 0$, and hence, $v(t) \geq J_0$ holds true for all times. This creates permanent misalignment between the two velocities at hand: v and $-v$.

So we can see that integrability of the kernel tail is the key. To obtain a general result, Cucker and Smale proposed to quantify connectivity strength of the flock as it evolves and proved that that strength is sustained by a non-integrable tail. This lies at the heart of their *spectral method* which we discuss next.

Let us consider the matrix $A = \{a_{ij}(t)\}_{i,j=1}^N \otimes \mathbb{I}_{n \times n}$, where $a_{ij} = \lambda\phi(\mathbf{x}_i(t), \mathbf{x}_j(t))$, $i \neq j$, and $a_{ij} = 0$, if $i = j$. Note that $a_{ij} \neq 0$ if and only if the corresponding agents are "connected" in the sense that they communicate through the influence kernel ϕ. By analogy with the graph theory, we call it the *adjacency matrix* of the flock. In fact, one can consider the actual adjacency matrix associated with the flock under the above connectivity definition: $\hat{A} = \{\hat{a}_{ij}(t)\}_{i,j=1}^N \otimes \mathbb{I}_{n \times n}$, where $\hat{a}_{ij} = 1$ if $a_{ij} \neq 0$, and 0 otherwise. Let $D = \text{diag}\{b_1, \ldots, b_N\} \otimes \mathbb{I}_{n \times n}$, where $b_i = \sum_j a_{ij}$, and $\hat{D} = \text{diag}\{\hat{b}_1, \ldots, \hat{b}_N\} \otimes \mathbb{I}_{n \times n}$, where $\hat{b}_i = \sum_j \hat{a}_{ij}$. Note that each $\hat{b}_i$ is precisely the degree of the vertex $\mathbf{x}_i$, i.e., the number of other agents to which it is connected.

In this notation, system (2.1) can now be written in terms of the grand velocity vector $\mathbf{V} = (\mathbf{v}_1, \ldots, \mathbf{v}_N)$, and the *Laplacian* associated to A, $L = D - A$, namely,

$$\frac{\mathrm{d}}{\mathrm{d}t}\mathbf{V} = -L\mathbf{V}. \tag{2.16}$$

The matrix L is nonnegative definite, and hence, the spectrum consists of a sequence $0 = \kappa_1 \leq \kappa_2 \leq \cdots \leq \kappa_N$. The average grand vector $\bar{\mathbf{V}} = (\bar{\mathbf{v}}, \ldots, \bar{\mathbf{v}})$ is obviously a member of the kernel of L; hence, $\kappa_1 = 0$. The next eigenvalue κ_2 is called the Fiedler number, although the classical Fiedler number is one associated to the Laplacian $\hat{L} = \hat{D} - \hat{A}$ in a similar way. We denote it by $\hat{\kappa}_2$. By the min-max theorem, κ_2 is given by

$$\kappa_2 = \min_{\sum_i \mathbf{v}_i = 0} \frac{\langle L\mathbf{V}, \mathbf{V} \rangle}{|\mathbf{V}|^2}.$$

A simple fact to verify is that $\kappa_2 \neq 0$ if and only if the graph is connected, and the relationship to $\hat{\kappa}_2$ is given by (see [34, Proposition 2]):

$$\kappa_2 \geq \hat{\kappa}_2 \min_{i,j:a_{ij} \neq 0} a_{ij}. \tag{2.17}$$

For this reason, we can call κ_2 a *weighted Fiedler number*, which captures not only algebraic connectivity of the flock as a graph but also the collective strength of the connection *weighted* by the kernel ϕ. With the use of the weighted Fiedler number $\kappa_2 = \kappa_2(t)$, which, let us recall, depends on time, we can measure alignment in (2.16) by writing the energy law as

$$\frac{d}{dt}|\mathbf{V} - \bar{\mathbf{V}}|^2 = -2\langle L(\mathbf{V} - \bar{\mathbf{V}}), (\mathbf{V} - \bar{\mathbf{V}})\rangle \le -2\kappa_2(t)|\mathbf{V} - \bar{\mathbf{V}}|^2.$$

Consequently,

$$|\mathbf{V}(t) - \bar{\mathbf{V}}| \le |\mathbf{V}_0 - \bar{\mathbf{V}}| \exp\left\{-\int_0^t \kappa_2(s)\, ds\right\}.$$

We can see that the divergence of the integral inside the curly brackets leads to alignment, which can be seen as a quantitative measure of connectivity as a function of time. Let us state it precisely.

Lemma 2.1 *If $\int_0^\infty \kappa_2(s)\, ds = \infty$, then the flock aligns.*

At the center of the original result of Cucker and Smale was a statement that for kernels of the type (1.3), the integral of the weighted Fiedler number is indeed divergent. Here, clearly, the flock remains always algebraically connected; thus, $\hat{\kappa}_2 = N$, and one can appeal directly to (2.17) to restate the problem in terms of control on the decay of the adjacency matrix. The original theorem of Cucker and Smale [33, 34] is the following:

Theorem 2.1 (Cucker-Smale) *Let $\phi(r) = \frac{1}{(1+r^2)^{\beta/2}}$. Then every solution aligns exponentially and flocks strongly for $\beta \le 1$ and conditionally if $\beta > 1$.*

In Sect. 2.4, this result will be proved using a Lyapunov function approach, which paves the way to extensions into kinetic and hydrodynamic systems. We will also state a sharp condition that ensures alignment even for the case of no heavy tail. Let us not, however, underestimate the spectral method as in some situations it is the only one available if no specific structural information is known about the kernel.

2.4   Alignment with Heavy Tail Communication

The main goal in this section will be to prove the Cucker-Smale Theorem 2.1 in the more general settings of a convolution-type heavy tail kernel using a method based on the maximum principle. We will see that the argument is rather versatile and is easily adaptable to other nonsymmetric kernels. So we consider the classical Cucker-Smale system:

$$\begin{cases} \dot{\mathbf{x}}_i = \mathbf{v}_i, \\ \dot{\mathbf{v}}_i = \lambda \sum_{k=1}^{N} m_k \phi(\mathbf{x}_i - \mathbf{x}_k)(\mathbf{v}_k - \mathbf{v}_i), \end{cases} \qquad (\mathbf{x}_i, \mathbf{v}_i) \in \mathbb{R}^n \times \mathbb{R}^n. \qquad (2.18)$$

We assume that ϕ is monotonely decreasing and everywhere positive. It will be useful to use sometimes the following shortcut notation:

$$\mathbf{x}_{ij} = \mathbf{x}_i - \mathbf{x}_j, \quad \mathbf{v}_{ij} = \mathbf{v}_i - \mathbf{v}_j, \quad \phi_{ij} = \phi(\mathbf{x}_i - \mathbf{x}_j), \quad \text{etc.}$$

Let us also consider the amplitude and flock diameter:

$$\mathcal{D} = \max_{i,j} |\mathbf{x}_{ij}|, \quad \mathcal{A} = \max_{i,j} |\mathbf{v}_{ij}|. \qquad (2.19)$$

Theorem 2.2 *Suppose ϕ is decreasing and positive and satisfies with initial data $\mathcal{D}_0, \mathcal{A}_0$,*

$$\int_{\mathcal{D}_0}^{\infty} \phi(r) \, dr > \frac{\mathcal{A}_0}{\lambda M}. \qquad (2.20)$$

Then the solution aligns and flocks exponentially fast:

$$\sup_{t \geq 0} \mathcal{D}(t) \leq \overline{\mathcal{D}}, \quad \mathcal{A}(t) \leq \mathcal{A}_0 e^{-t\lambda M \phi(\overline{\mathcal{D}})}. \qquad (2.21)$$

In particular, every solution flocks provided the kernel satisfies the heavy tail condition (2.6).

To make the proof perfectly rigorous, let us recall the classical Rademacher Lemma, which we will use throughout.

Suppose $f(x, t) : X \times \mathbb{R}_+ \to \mathbb{R}$ is a Lipschitz in time function uniformly in x, where X is an arbitrary index set. So $\exists L > 0$ such that for all t, s, x we have

$$|f(x, t) - f(x, s)| \leq L|t - s|.$$

Suppose that at any time t there is a point $x(t) \in X$ such that

$$f(x(t), t) = \sup_{x \in X} f(x, t) := F(t).$$

Note that $F(t)$ is a Lipschitz function with the same constant L. Indeed, let $t, s \in \mathbb{R}_+$ and $F(t) > F(s)$. Then

$$F(t) - F(s) = f(x(t), t) - f(x(t), s) + \underbrace{f(x(t), s) - f(x(s), s)}_{\leq 0}$$

$$\leq f(x(t), t) - f(x(t), s) \leq L|t - s|.$$

Consequently, F is absolutely continuous on any finite interval; that is, there exists $m \in L^\infty$ such that

$$F(t) - F(s) = \int_s^t m(\tau)d\tau, \qquad F' = m, \text{ a.e.}$$

Lemma 2.2 *If $f(x, \cdot)$ is differentiable everywhere in t for all $x \in X$, then $F'(t) = \partial_t f(x(t), t)$ holds at any point where F' exists.*

Indeed, computing the one-sided derivative from the right, we have

$$F'(t) = \lim_{h \to 0+} \frac{f(x(t+h), t+h) - f(x(t), t+h) + f(x(t), t+h) - f(x(t), t)}{h}$$

$$\geq \lim_{h \to 0+} \frac{f(x(t), t+h) - f(x(t), t)}{h} = \partial_t f(x(t), t).$$

Taking $h < 0$ proves the opposite inequality.

Proof (Proof of Theorem 2.2) Let us represent $\mathcal{A}$ as

$$\mathcal{A} = \max_{|\ell|=1, i, j} \ell(\mathbf{v}_{ij}). \tag{2.22}$$

Here, ℓ is considered to be an element of the dual space $(\mathbb{R}^n)^*$. Note that the maximum is taken over a fixed compact set not changing in time. So we can pick ℓ and i, j at each time t for which the maximum is achieved. Using Rademacher's lemma and the velocity equation, we obtain

$$\frac{d}{dt} \ell(\mathbf{v}_{ij}) = \lambda \sum_{k=1}^N m_k \phi_{ik} \ell(\mathbf{v}_{ki}) + m_k \phi_{jk} \ell(\mathbf{v}_{jk}). \tag{2.23}$$

Now notice that $\ell(\mathbf{v}_{ki}) \leq 0$ and $\ell(\mathbf{v}_{jk}) \leq 0$. One can see this by adding and subtracting $\mathbf{v}_j$ in the first case and $\mathbf{v}_i$ in the second and using maximality of $\ell(\mathbf{v}_{ij})$. So then we can pull out the minimal value of the kernel in both sums:

$$\frac{d}{dt} \ell(\mathbf{v}_{ij}) \leq -\lambda \phi(\mathcal{D}) \sum_{k=1}^N m_k \ell(\mathbf{v}_{ij}) = -\lambda M \phi(\mathcal{D}) \ell(\mathbf{v}_{ij}).$$

So we obtain

$$\begin{cases} \dfrac{d}{dt}\mathcal{A} \le -\lambda M\phi(\mathcal{D})\mathcal{A} \\[2mm] \dfrac{d}{dt}\mathcal{D} \le \mathcal{A}. \end{cases} \tag{2.24}$$

This system of ordinary differential inequalities (ODIs) has a decreasing Lyapunov function given by $L = \mathcal{A} + \lambda M \int_0^{\mathcal{D}} \phi(r)\,dr$. This, in particular, implies that

$$\lambda M \int_0^{\mathcal{D}(t)} \phi(r)\,dr \le \mathcal{A}_0 + \lambda M \int_0^{\mathcal{D}_0} \phi(r)\,dr, \quad \forall t > 0.$$

Consequently, $\mathcal{D}(t) \le \overline{\mathcal{D}}$, where $\overline{\mathcal{D}}$ is obtained from the equation

$$\lambda M \int_{\mathcal{D}_0}^{\overline{\mathcal{D}}} \phi(r)\,dr = \mathcal{A}_0, \tag{2.25}$$

which is guaranteed to have a finite solution, thanks to (2.20). Then,

$$\dot{\mathcal{A}} \le -\lambda M\phi(\overline{\mathcal{D}})\mathcal{A}$$

and the theorem follows. $\qquad\qquad\qquad\qquad\qquad\qquad\qquad\qquad\qquad\qquad\square$

Solving Equation (2.25) allows one to provide explicit decay rates for solutions of (2.24) for some kernels. In particular, for the classical Cucker-Smale kernel

$$\phi(r) = \frac{1}{(1+r^2)^{\frac{\beta}{2}}},$$

one obtains

$$\overline{\mathcal{D}} \le \left(\left[\frac{1-\beta}{\lambda M}\mathcal{A}_0 + (1+\mathcal{D}_0^2)^{\frac{1-\beta}{2}} \right]^{\frac{2}{1-\beta}} - 1 \right)^{\frac{1}{2}}, \quad \beta < 1,$$

$$\text{rate} = \frac{\lambda M}{\left[\frac{1-\beta}{\lambda M}\mathcal{A}_0 + (1+\mathcal{D}_0^2)^{\frac{1-\beta}{2}} \right]^{\frac{\beta}{1-\beta}}}, \tag{2.26}$$

(here one replaces ϕ with a smaller but explicitly integrable kernel $\dfrac{r}{(1+r^2)^{\frac{\beta+1}{2}}}$), and

$$\overline{\mathcal{D}} \le \left(e^{\frac{2}{\lambda M}\mathcal{A}_0}(1+\mathcal{D}_0^2) - 1 \right)^{\frac{1}{2}}, \quad \text{rate} = \frac{\lambda M}{e^{\frac{\mathcal{A}_0}{\lambda M}}(1+\mathcal{D}_0^2)^{\frac{1}{2}}}, \quad \beta = 1. \tag{2.27}$$

The proof of Theorem 2.2 is rather flexible and can be adapted to other models, including those with nonsymmetric communication; see Sect. 2.8.

Remark 2.1 It is interesting to note that the argument presented above is independent of any particular choice of a norm $|\cdot|$ on $\mathbb{R}^n$, as long as the unit functionals in (2.22) are chosen in the ball of the corresponding dual space. The basic system of inequalities (2.24) would still hold true with exact same coefficients, independent of the chosen norm. This remark applies to many of the subsequent results below.

2.5 Stability

Suppose now we have two close initial conditions for the same flock with the same momentum:

$$\mathbf{V} = (\mathbf{v}_i)_{i=1}^N, \ \mathbf{X} = (\mathbf{x}_i)_{i=1}^N, \quad \widetilde{\mathbf{V}} = (\tilde{\mathbf{v}}_i)_{i=1}^N, \ \widetilde{\mathbf{X}} = (\tilde{\mathbf{x}}_i)_{i=1}^N,$$

and $\bar{\mathbf{v}} = \bar{\tilde{\mathbf{v}}}$. We show that if these parameters are close initially, then they will remain close uniformly for all time. To formulate this precisely, let us introduce the measure of distance between flocks:

$$\|\mathbf{X} - \widetilde{\mathbf{X}}\| = \max_i |\mathbf{x}_i - \tilde{\mathbf{x}}_i|, \quad \|\mathbf{V} - \widetilde{\mathbf{V}}\| = \max_i |\mathbf{v}_i - \tilde{\mathbf{v}}_i|. \tag{2.28}$$

Much in the spirit of the previous proofs, let us derive a system of ODIs for these two quantities. First, we clearly have

$$\frac{\mathrm{d}}{\mathrm{d}t}\|\mathbf{X} - \widetilde{\mathbf{X}}\| \le \|\mathbf{V} - \widetilde{\mathbf{V}}\|.$$

Second, let us represent

$$\|\mathbf{V} - \widetilde{\mathbf{V}}\| = \max_{i,|\ell|=1} \ell(\mathbf{v}_i - \tilde{\mathbf{v}}_i),$$

and by Rademacher's lemma apply the derivative in time with a maximizing pair i, ℓ:

$$\frac{\mathrm{d}}{\mathrm{d}t}\ell(\mathbf{v}_i - \tilde{\mathbf{v}}_i) = \lambda \sum_{k=1}^{N} m_k \phi(\mathbf{x}_{ik})\ell(\mathbf{v}_{ki}) - m_k \phi(\tilde{\mathbf{x}}_{ik})\ell(\tilde{\mathbf{v}}_{ki})$$

$$= \lambda \sum_{k=1}^{N} m_k (\phi(\mathbf{x}_{ik}) - \phi(\tilde{\mathbf{x}}_{ik}))\ell(\mathbf{v}_{ki})$$

$$+ \lambda \sum_{k=1}^{N} m_k \phi(\tilde{\mathbf{x}}_{ik})[\ell(\mathbf{v}_k - \tilde{\mathbf{v}}_k) - \ell(\mathbf{v}_i - \tilde{\mathbf{v}}_i)]$$

$$\leq 2M\lambda \|\nabla\phi\|_\infty \mathcal{A}\|\mathbf{X} - \tilde{\mathbf{X}}\| + \lambda\phi(\overline{\overline{\mathcal{D}}}) \sum_{k=1}^{N} m_k[\ell(\mathbf{v}_k - \tilde{\mathbf{v}}_k) - \ell(\mathbf{v}_i - \tilde{\mathbf{v}}_i)]$$

$$= 2M\lambda \|\nabla\phi\|_\infty \mathcal{A}_0 e^{-t\lambda M\phi(\overline{\mathcal{D}})}\|\mathbf{X} - \tilde{\mathbf{X}}\| - M\lambda\phi(\overline{\overline{\mathcal{D}}})\|\mathbf{V} - \tilde{\mathbf{V}}\|.$$

Let us recall that the bounds on the diameters ultimately depend on the initial data via (2.25). If the initial flock $\mathbf{X}$, $\mathbf{V}$ is known and the perturbation $\tilde{\mathbf{X}}$, $\tilde{\mathbf{V}}$ is relatively small, then those diameters can be quantified by the initial values of $\mathcal{D}_0$ and $\mathcal{A}_0$. So let us simply denote

$$\gamma = \min\{\phi(\overline{\mathcal{D}}), \phi(\overline{\overline{\mathcal{D}}})\}.$$

We have obtained the system

$$\begin{cases} \dfrac{\mathrm{d}}{\mathrm{d}t}\|\mathbf{V} - \tilde{\mathbf{V}}\| \leq 2M\lambda \|\nabla\phi\|_\infty \mathcal{A}_0 e^{-t\lambda\gamma M}\|\mathbf{X} - \tilde{\mathbf{X}}\| - \lambda\gamma M\|\mathbf{V} - \tilde{\mathbf{V}}\| \\[2mm] \dfrac{\mathrm{d}}{\mathrm{d}t}\|\mathbf{X} - \tilde{\mathbf{X}}\| \leq \|\mathbf{V} - \tilde{\mathbf{V}}\|. \end{cases} \qquad (2.29)$$

Let us simply rewrite it as

$$x' \leq v, \quad v' \leq ae^{-bt}x - bv. \qquad (2.30)$$

It is elementary to obtain a bound on solutions. Indeed, denoting $w = ve^{bt}$, we obtain

$$x' \leq we^{-bt}, \quad w' \leq ax.$$

Multiplying by suitable factors to equalize the right-hand sides, we obtain

$$\frac{\mathrm{d}}{\mathrm{d}t}(ax^2 + e^{-bt}w^2) \leq 4axwe^{-bt} \leq 2e^{-bt/2}\sqrt{a}(ax^2 + e^{-bt}w^2).$$

This immediately implies

$$ax^2 + e^{bt}v^2 \le e^{\frac{4\sqrt{a}}{b}}(ax_0^2 + v_0^2).$$

We can read off bounds for each parameter individually:

$$x \le \frac{1}{\sqrt{a}} e^{\frac{2\sqrt{a}}{b}} \sqrt{ax_0^2 + v_0^2}, \quad v \le e^{-bt/2 + \frac{2\sqrt{a}}{b}} \sqrt{ax_0^2 + v_0^2}. \tag{2.31}$$

Theorem 2.3 *The following bound holds for any pair of solutions to* (2.18) *with the same momentum:*

$$a\|\mathbf{X} - \tilde{\mathbf{X}}\|^2 + e^{bt}\|\mathbf{V} - \tilde{\mathbf{V}}\|^2 \le e^{\frac{4\sqrt{a}}{b}}(a\|\mathbf{X}_0 - \tilde{\mathbf{X}}_0\|^2 + \|\mathbf{V}_0 - \tilde{\mathbf{V}}_0\|^2), \tag{2.32}$$

where $a = 2M\lambda\|\nabla\phi\|_\infty \mathcal{A}_0$ *and* $b = \lambda M \min\{\phi(\overline{\mathcal{D}}), \phi(\overline{\mathcal{D}})\}.$

2.6   Singular Kernels and the Issue of Collisions

Before we proceed further, let us bring into consideration singular kernels (2.5), which will play an important role in alleviating several issues related to dynamics with smooth communication. One such issue is collision of agents. While at the discrete level it presents no difficulty from the point of view of regularity of the ODE, at the macroscopic level this results in finite time blow-up, which will be discussed in Chap. 7. Let us start with an example.

Example 2.2 Let us assume that $\phi = 1$ in a neighborhood of 0. Let us arrange two agents $x = x_1 = -x_2$ with $0 < x(0) = \varepsilon \ll 1$. And let $v_1 = -v_2 < 0$ be very large. Clearly, $x(t)$ will remain in the same neighborhood of 0 where it has started, and the system reads

$$\frac{d}{dt}x = v, \quad \frac{d}{dt}v = -2v.$$

Solving it explicitly, we can see that the two agents will collide in finite time at the origin.

In 1D, one can in fact give a complete answer to the question which initial conditions lead to collision and which do not. To this end, let us assume that ϕ is a smooth radial convolution kernel and consider its primitive $\psi' = \phi$. Denote

$$f_i = v_i + \sum_{k=1}^{N} \psi(x_i - x_k)m_k. \tag{2.33}$$

All these quantities are conserved:

$$\frac{d}{dt} f_i = 0.$$

Collision can now be characterized entirely in terms of a collection of conservation laws $\{f_i\}_{i=1}^N$.

Theorem 2.4 *An initial non-collisional data*

$$x_1(0) < x_2(0) < \cdots < x_N(0)$$

remains non-collisional at all times if and only if

$$f_1 \leq f_2 \leq \cdots \leq f_N.$$

Moreover, if

$$f_1 < f_2 < \cdots < f_N,$$

then the distance between agents remains bounded away from zero:

$$\inf_{i,j,t>0} |x_i(t) - x_j(t)| > 0.$$

Proof Simply observe the identity

$$\frac{d}{dt}(x_i - x_j) = f_i - f_j - \sum_{k=1}^N \int_{x_i}^{x_j} \phi(y - x_k) m_k \, dy.$$

We can see that if f is monotonically increasing, then

$$\frac{d}{dt}(x_i - x_j) \geq f_{ij} - \|\phi\|_\infty M(x_i - x_j),$$

where $f_{ij} \geq 0$, for $i > j$. By Grönwall's lemma the order of the agents will be preserved, and

$$x_i - x_j \geq e^{-\|\phi\|_\infty M t}(x_i(0) - x_j(0)) + \frac{f_{ij}}{\|\phi\|_\infty M}(1 - e^{-\|\phi\|_\infty M t}).$$

This proves the theorem in the positive direction. Conversely, if for some i, $f_i > f_{i+1}$, then

$$\frac{d}{dt}(x_{i+1} - x_i) \leq f_{i+1} - f_i,$$

and hence, the two agents will collapse in finite time.   □

If the kernel ϕ is singular, it greatly enhances alignment forces at close range. So if two agents are on a collision course toward each other, the alignment will help to slow down the forward momentum and possibly avoid the collision. Just how much singularity is needed to make this work is illustrated by the following example.

Example 2.3 Let the kernel be given by (2.5) and let us consider the same setup as previously. Then we obtain the system of ODEs

$$\frac{d}{dt}x = v, \quad \frac{d}{dt}v = -2\frac{v}{x^\beta}.$$

This system has a conservation law provided $\beta < 1$: $v + \frac{2x^{1-\beta}}{1-\beta} = C_0$. So if initially $C_0 \ll 0$, then $v < C_0 \ll 0$ as well. This means that x will reach the origin in finite time.

This example demonstrates that the threshold singularity necessary to prevent collisions must be non-integrable. The next result shows that this condition is sufficient as well, and it holds in any dimensions.

Theorem 2.5 *Under the short range heavy tail condition* (2.7), *the flock experiences no collisions between agents for any non-collisional initial datum. Consequently, any non-collisional initial datum gives rise to a unique global solution to the Cucker-Smale system* (2.1).

In view of global existence and the absence of collisions, the content of Theorem 2.2 holds true as stated provided the kernel is singular in the sense of condition (2.7).

Proof For given non-collisional initial data $(\mathbf{x}_i, \mathbf{v}_i)_i$, let us assume that collision occurs first time at $t = T^*$. Denote by $I^* \subset \{1, \ldots, N\}$ the indexes of agents that collided at one point in space—note that this may not be a unique collection. Consequently, there is a $\delta > 0$ such that $|\mathbf{x}_{ik}(t)| \geq \delta$ for all $i \in I^*$ and $k \in \Omega \backslash I^*$. Let us denote

$$\mathcal{D}^*(t) = \max_{i,j \in I^*} |\mathbf{x}_{ij}(t)|, \quad \mathcal{A}^*(t) = \max_{i,j \in I^*} |\mathbf{v}_{ij}(t)|, \quad t < T^*.$$

Directly from the characteristic equation, we obtain $\frac{d}{dt}|\mathcal{D}^*| \leq \mathcal{A}^*$, and hence,

$$-\frac{d}{dt}\mathcal{D}^* \leq \mathcal{A}^*. \tag{2.34}$$

For velocity amplitude $\mathcal{A}^*$, we obtain, using a maximizing triple $\ell \in (\mathbb{R}^n)^*$, $i, j \in I^*$:

$$\frac{\mathrm{d}}{\mathrm{d}t}\mathcal{A}^* = \sum_{k=1}^{N} m_k \phi_{ik} \ell(\mathbf{v}_{ki}) - m_k \phi_{kj} \ell(\mathbf{v}_{kj}) = \sum_{k \in I^*} m_k \phi_{ik} \ell(\mathbf{v}_{kj} - \mathbf{v}_{ij})$$

$$+ m_k \phi_{kj} \ell(-\mathbf{v}_{ki} - \mathbf{v}_{ij}) + \sum_{k \notin I^*} m_k \phi_{ik} \ell(\mathbf{v}_{ki}) - m_k \phi_{kj} \ell(\mathbf{v}_{kj}).$$

In the first sum, we notice that all terms are negative, so we can pull out the minimal value of the kernel, which is $\phi(\mathcal{D}^*)$, and the rest add up to a constant multiple of $-\mathcal{A}^*$. In the second sum, all the distances $|\mathbf{x}_{ik}|$, $|\mathbf{x}_{jk}|$ stay away from zero up to T^*. So the kernels and the whole sum remain bounded. In summary, we obtain

$$\frac{\mathrm{d}}{\mathrm{d}t}\mathcal{A}^* \leq C_1 - C_2 \phi(\mathcal{D}^*)\mathcal{A}^*.$$

Considering the energy functional

$$E(t) = \mathcal{A}^*(t) + C_2 \int_{\mathcal{D}^*(t)}^{1} \phi(r)\,\mathrm{d}r,$$

we readily find that $\dfrac{\mathrm{d}}{\mathrm{d}t}E \leq C_1$; hence, E remains bounded up to the critical time. This means that $\mathcal{D}^*(t)$ cannot approach zero value.

The global existence part is now a routine application of the Picard iteration and the standard continuation argument.   □

When the kernel has a more explicitly defined power law singularity, such as

$$\phi(r) \gtrsim \frac{\mathbb{1}_{r<r_0}}{r^\beta}, \quad \text{for } \beta \geq 2, \tag{2.35}$$

a more quantitative estimate on the minimal distance between agents can be derived. Indeed, let us consider the following *collision functional*:

$$\mathcal{C} = \begin{cases} \dfrac{1}{N^2} \displaystyle\sum_{i,j=1}^{N} \dfrac{1}{(|\mathbf{x}_{ij}| \wedge r_0)^{\beta-2}}, & \beta > 2, \\[4mm] \dfrac{1}{N^2} \displaystyle\sum_{i,j=1}^{N} \ln(|\mathbf{x}_{ij}| \wedge r_0), & \beta = 2. \end{cases} \tag{2.36}$$

For $\beta > 2$, we have for the derivative

$$\frac{dC}{dt} = \frac{(2-\beta)}{N^2} \sum_{i,j=1}^{N} \frac{\frac{d}{dt}\left(|\mathbf{x}_{ij}| \wedge r_0\right)}{(|\mathbf{x}_{ij}| \wedge r_0)^{\beta-1}} \leq \frac{|\beta-2|}{N^2} \sum_{i,j=1}^{N} \frac{|\mathbf{v}_{ij}| \, \mathbb{1}_{|\mathbf{x}_{ij}|<r_0}}{(|\mathbf{x}_{ij}| \wedge r_0)^{\beta-1}}$$

$$\leq |\beta-2| \left(\frac{1}{N^2} \sum_{i,j=1}^{N} \mathbf{v}_{ij}^2 \frac{1}{|\mathbf{x}_{ij}|^\beta} \mathbb{1}_{|\mathbf{x}_{ij}|<r_0} \right)^{1/2}$$

$$\times \left(\frac{1}{N^2} \sum_{i,j=1}^{N} |\mathbf{x}_{ij}|^{2-\beta} \mathbb{1}_{|\mathbf{x}_{ij}|<r_0} \right)^{1/2}$$

$$\leq C\sqrt{\mathcal{I}_2}\sqrt{C}.$$

This implies that

$$\sqrt{C(t)} \leq \sqrt{C(0)} + C \int_0^t \sqrt{\mathcal{I}_2(s)}\, ds, \tag{2.37}$$

and recalling that $\mathcal{I}_2$ is integrable on $\mathbb{R}^+$, we conclude that

$$C(t) \lesssim t. \tag{2.38}$$

For $\beta = 2$, a similar computation gives $\frac{d}{dt}C \leq C\sqrt{\mathcal{I}_2}$; hence, $C(t) \lesssim \sqrt{t}$. We thus arrive at the following bounds:

$$|\mathbf{x}_{ij}(t)| \geq \begin{cases} \dfrac{c}{t^{\frac{1}{\beta-2}}}, & \beta > 2, \\[2ex] ce^{-C\sqrt{t}}, & \beta = 2. \end{cases} \tag{2.39}$$

2.7 Degenerate Communication: Corrector Method

Non-degenerate communication is not always realistic if we think of biological systems such as flocks. Indeed, every bird has only limited sensory abilities. In addition, in close range, the alignment force may not even be the most dominant as it appears, for example, in 3-zone models, which we will discuss in Chap. 3. The case when the kernel is degenerate, i.e., it vanishes at some points, presents obvious difficulty with the implementation of the Lyapunov function approach. The main difficulty is of course lack of any coercivity in the energy law:

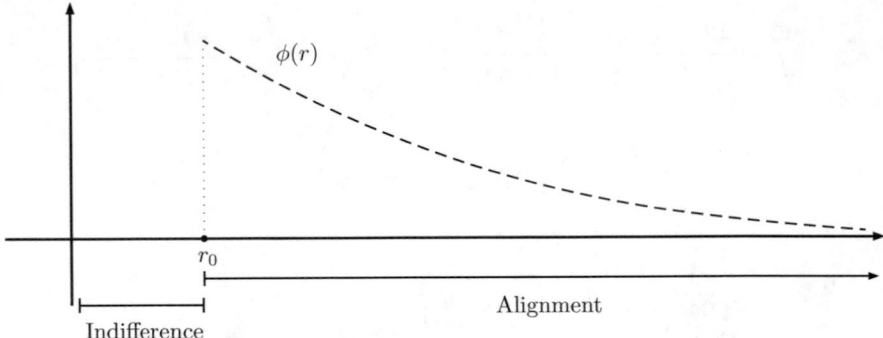

Fig. 2.3 Example of degenerate communication at close range

$$\frac{\mathrm{d}}{\mathrm{d}t}\mathcal{V}_2 = -2\mathcal{I}_2. \tag{2.40}$$

If the kernel is non-degenerate only in a bounded range $r \leq r_0$, for example, as depicted in Fig. 2.3, then there still exists a dynamical mechanism to restore communication. Indeed, if two agents $\mathbf{x}_i$, $\mathbf{x}_j$ are not yet aligned, then their velocities are not the same. Hence, they would tend to separate from each other, and provided no other influence changes their course, they eventually reach the communication range $r > r_0$. In this section, we present a method introduced in [38] which allows to partially fix the lack of coercivity by introducing a proper *energy corrector*. The method utilizes higher-order fluctuation functionals given by

$$\mathcal{V}_p = \frac{1}{pN^2} \sum_{i,j=1}^{N} |\mathbf{v}_i - \mathbf{v}_j|^p, \quad p \geq 1,$$

$$\mathcal{I}_p = \frac{1}{pN^2} \sum_{i,j=1}^{N} \phi(\mathbf{x}_i, \mathbf{x}_j)|\mathbf{v}_i - \mathbf{v}_j|^p. \tag{2.41}$$

We observe that the $\mathcal{V}_p$'s are non-increasing. Indeed,

$$\frac{\mathrm{d}}{\mathrm{d}t}\mathcal{V}_p = \frac{1}{N^3} \sum_{i,j,k} |\mathbf{v}_{ij}|^{p-2}\mathbf{v}_{ij} \cdot (\mathbf{v}_{ki}\phi_{ki} - \mathbf{v}_{kj}\phi_{kj}) = \frac{2}{N^3} \sum_{i,j,k} |\mathbf{v}_{ij}|^{p-2}\mathbf{v}_{ij} \cdot \mathbf{v}_{ki}\phi_{ki}$$

$$= \frac{1}{N^3} \sum_{i,j,k} (|\mathbf{v}_{ij}|^{p-2}\mathbf{v}_{ij} - |\mathbf{v}_{kj}|^{p-2}\mathbf{v}_{kj}) \cdot \mathbf{v}_{ki}\phi_{ki}$$

$$= \frac{1}{N^3} \sum_{i,j,k} (|\mathbf{v}_{ij}|^{p-2}\mathbf{v}_{ij} - |\mathbf{v}_{kj}|^{p-2}\mathbf{v}_{kj}) \cdot (\mathbf{v}_{kj} - \mathbf{v}_{ij})\phi_{ki},$$

with the convention that $|\mathbf{v}_{ij}|^{p-2}\mathbf{v}_{ij} = 0$ if $i = j$. The right-hand side is nonpositive, thanks to the elementary inequality:

$$\left(|a|^{p-2}a - |b|^{p-2}b\right) \cdot (a - b) \geq 0.$$

The two special cases, $i = j$ and $k = j$, produce the term $-|\mathbf{v}_{ik}|^p - |\mathbf{v}_{ij}|^p$. So in general, we have an N-dependent inequality:

$$\frac{d}{dt}\mathcal{V}_p \leq -\frac{p}{N}\mathcal{I}_p, \quad p \geq 1. \tag{2.42}$$

For the case $p = 2$, the energy law is of course N-independent (2.40).

Theorem 2.6 *Suppose the kernel $\phi \geq 0$ is either smooth or satisfies* (2.7). *Suppose also that it dominates a monotone heavy tail: there exists a non-increasing $\Phi(r)$, such that for some $r_0 > 0$,*

$$\phi(r) \geq \Phi(r), \ \forall r > r_0, \quad and \quad \int_{r_0}^{\infty} \Phi(r)\,dr = \infty. \tag{2.43}$$

Then

(i) Any solution to the discrete system (2.1) aligns: $\mathcal{V}_2(t) \leq \frac{C_N}{t}$, and $\mathcal{D}(t) \leq \overline{\mathcal{D}}_N$, with constants depending on N.

(ii) If ϕ is smooth, then any solution to the discrete system (2.1) aligns: $\mathcal{V}_4 \to 0$, with a rate independent of N.

The advantage of (i) over (ii) is that it gives a faster, although N-dependent, rate as well as flocking. It also holds for singular kernels satisfying the short range singularity condition (2.7). However, it is not extendable to the macroscopic or kinetic case, while (ii) is.

Proof (Proof of Theorem 2.6 (i)) We start by defining the following corrector:

$$\mathcal{G} = \frac{1}{N^2} \sum_{i,j=1}^{N} |\mathbf{v}_{ij}|\psi(d_{ij})\chi(|\mathbf{x}_{ij}|),$$

where d_{ij} is a longitudinal displacement function defined by

$$d_{ij} = -\mathbf{x}_{ij} \cdot \frac{\mathbf{v}_{ij}}{|\mathbf{v}_{ij}|}, \tag{2.44}$$

and the two auxiliary functions $\chi : \mathbb{R}^+ \mapsto \mathbb{R}^+$ and $\psi : \mathbb{R} \mapsto \mathbb{R}^+$ are defined by

$$\chi(r) = \begin{cases} 1, & r < r_0, \\ 2 - \frac{r}{r_0}, & r_0 \leq r \leq 2r_0, \quad \text{and} \quad \psi(d) = \begin{cases} 0, & d < -r_0, \\ d + r_0, & |d| \leq r_0, \\ 2r_0, & d > r_0. \end{cases} \\ 0, & r > 2r_0 \end{cases}$$

Let us compute the derivative of the corrector:

$$\frac{d}{dt}\mathcal{G} = -\frac{1}{N^2}\sum_{i,j=1}^{N}|\mathbf{v}_{ij}|^2 \mathbb{1}_{|d_{ij}|<r_0}\chi(|\mathbf{x}_{ij}|) + \mathcal{R}_1 + \mathcal{R}_2 + \mathcal{R}_3, \tag{2.45}$$

where

$$\mathcal{R}_1 = \frac{2}{N^3}\sum_{i,j,k=1}^{N}\frac{\mathbf{v}_{ij}}{|\mathbf{v}_{ij}|}\cdot\mathbf{v}_{ki}\phi(\mathbf{x}_{ik})\psi(d_{ij})\chi(|\mathbf{x}_{ij}|),$$

$$\mathcal{R}_2 = -\frac{2}{N^3}\sum_{i,j,k=1}^{N}\mathbf{x}_{ij}\cdot\left(\mathbb{I} - \frac{\mathbf{v}_{ij}\otimes\mathbf{v}_{ij}}{|\mathbf{v}_{ij}|^2}\right)\mathbf{v}_{ki}\phi(\mathbf{x}_{ik})\mathbb{1}_{|d_{ij}|<r_0}\chi(|\mathbf{x}_{ij}|),$$

$$\mathcal{R}_3 = \frac{1}{N^2}\sum_{i,j=1}^{N}|\mathbf{v}_{ij}|\psi(d_{ij})\chi'(|\mathbf{x}_{ij}|)\frac{\mathbf{x}_{ij}}{|\mathbf{x}_{ij}|}\cdot\mathbf{v}_{ij}.$$

The first term on the right-hand side of (2.45) patches up communication where it was originally missing—at the close range. Let us address it first. Without loss of generality, we can assume that Φ is a bounded decreasing function on $\mathbb{R}^+$. Hence,

$$\mathbb{1}_{r<r_0} + \phi(r) \geq c\Phi(r),$$

for all $r > 0$ and some $c > 0$. Using $\mathbb{1}_{|d_{ij}|<r_0} \geq \mathbb{1}_{|\mathbf{x}_{ij}|<r_0}$, we have

$$-\frac{1}{N^2}\sum_{i,j}|\mathbf{v}_{ij}|^2\mathbb{1}_{|d_{ij}|<r_0}\chi(|\mathbf{x}_{ij}|) \leq -\frac{1}{N^2}\sum_{i,j}|\mathbf{v}_{ij}|^2\mathbb{1}_{|\mathbf{x}_{ij}|<r_0}$$

$$= -\frac{1}{N^2}\sum_{i,j}|\mathbf{v}_{ij}|^2(\mathbb{1}_{|\mathbf{x}_{ij}|<r_0} + \phi(\mathbf{x}_{ij})) + \frac{1}{N^2}\sum_{i,j}|\mathbf{v}_{ij}|^2\phi(\mathbf{x}_{ij})$$

$$\leq -c\Phi(\mathcal{D})\mathcal{V}_2 + \mathcal{I}_2.$$

Proceeding to the error terms, by construction of the auxiliary functions, we have

$$|\mathcal{R}_1|, |\mathcal{R}_2| \lesssim \mathcal{I}_1,$$

and

$$\mathcal{R}_3 \le \frac{1}{N^2} \sum_{i,j} |\mathbf{v}_{ij}|^2 |\chi'(|\mathbf{x}_{ij}|)| \le \frac{1}{N^2} \sum_{i,j} |\mathbf{v}_{ij}|^2 \mathbb{1}_{r_0 < |\mathbf{x}_{ij}| < 2r_0}$$

$$\lesssim \frac{1}{N^2} \sum_{i,j} |\mathbf{v}_{ij}|^2 \phi(\mathbf{x}_{ij}) = \mathcal{I}_2.$$

Hence, with constants a, b, c only depending on ϕ, we obtain

$$\frac{\mathrm{d}}{\mathrm{d}t} \mathcal{G} \le -c\Phi(\mathcal{D})\mathcal{V}_2 + a\mathcal{I}_2 + b\mathcal{I}_1.$$

We can now define a new Lyapunov functional:

$$\mathcal{L} = \mathcal{G} + a\mathcal{V}_2 + bN\mathcal{V}_1, \qquad \frac{\mathrm{d}}{\mathrm{d}t} \mathcal{L} \le -c\Phi(\mathcal{D})\mathcal{V}_2. \tag{2.46}$$

Using that $\dfrac{\mathrm{d}}{\mathrm{d}t} \mathcal{D} \le C_N \sqrt{\mathcal{V}_2}$, we find another Lyapunov functional:

$$\tilde{\mathcal{L}} = \mathcal{L} + \frac{c}{C_N} \sqrt{\mathcal{V}_2} \int_0^{\mathcal{D}} \Phi(r)\,\mathrm{d}r.$$

Consequently, thanks to the heavy tail of Φ, we have flocking $\mathcal{D}(t) \le \overline{\mathcal{D}}_N$ for all time and for some N-dependent $\overline{\mathcal{D}}_N$. Returning to (2.46), we conclude that

$$A := \int_0^\infty \mathcal{V}_2(t)\,\mathrm{d}t < \infty.$$

So on every time interval $[T, e^A T]$, there is a t such that $\mathcal{V}_2(t) \le \frac{1}{t}$. By monotonicity, this implies a similar bound for all t.   □

Proof (Proof of Theorem 2.6 (ii)) To achieve an N-independent, although slower, rate, we consider a third-order corrector given by

$$\mathcal{G}_3 = \frac{1}{N^2} \sum_{i,j} |\mathbf{v}_{ij}|^3 \psi(d_{ij}) \chi(|\mathbf{x}_{ij}|),$$

with ψ and χ defined previously. In this case,

$$\frac{\mathrm{d}}{\mathrm{d}t} \mathcal{G}_3 = -\frac{1}{N^2} \sum_{i,j=1}^{N} |\mathbf{v}_{ij}|^4 \mathbb{1}_{|d_{ij}| < r_0} \chi(|\mathbf{x}_{ij}|) + \mathcal{R}_1 + \mathcal{R}_2 + \mathcal{R}_3,$$

with

$$\mathcal{R}_1 = \frac{6}{N^3} \sum_{i,j,k=1}^{N} |\mathbf{v}_{ij}| \mathbf{v}_{ij} \cdot \mathbf{v}_{ki} \, \phi_{ik} \, \psi(d_{ij}) \chi(|\mathbf{x}_{ij}|),$$

$$\mathcal{R}_2 = \frac{-2}{N^3} \sum_{i,j,k=1}^{N} |\mathbf{v}_{ij}|^2 \, \mathbf{x}_{ij} \cdot \left(\mathbb{I} - \frac{\mathbf{v}_{ij} \otimes \mathbf{v}_{ij}}{|\mathbf{v}_{ij}|^2} \right) \mathbf{v}_{ki} \phi_{ki} \, \mathbb{1}_{|d_{ij}|<r_0} \chi(|\mathbf{x}_{ij}|)$$

$$\mathcal{R}_3 = \frac{1}{N^2} \sum_{i,j=1}^{N} |\mathbf{v}_{ij}|^3 \psi(d_{ij}) \chi'(|\mathbf{x}_{ij}|) \frac{\mathbf{x}_{ij}}{|\mathbf{x}_{ij}|} \cdot \mathbf{v}_{ij}.$$

The gain term is estimated as before with the use of a priori uniform bound on velocities $|\mathbf{v}_i(t)| \le \max_i |\mathbf{v}_i(0)| = C$:

$$-\frac{1}{N^2} \sum_{i,j=1}^{N} |\mathbf{v}_{ij}|^4 \mathbb{1}_{|d_{ij}|<r_0} \chi(|\mathbf{x}_{ij}|) \le -c\Phi(\mathcal{D})\mathcal{V}_4 + C^2 \mathcal{I}_2. \tag{2.47}$$

Next,

$$\mathcal{R}_3 \lesssim \frac{1}{N^2} \sum_{i,j} |\mathbf{v}_{ij}|^3 \phi_{ij} \lesssim \mathcal{I}_2.$$

By Young's inequality for a small $\epsilon > 0$ to be specified later,

$$\mathcal{R}_2 \lesssim \frac{\varepsilon}{N^2} \sum_{i,j=1}^{N} |\mathbf{v}_{ij}|^4 |\mathbf{x}_{ij}|^2 \chi(|\mathbf{x}_{ij}|)^2 + \frac{\|\phi\|_\infty}{\varepsilon} \frac{1}{N^2} \sum_{i,k=1}^{N} |\mathbf{v}_{ki}|^2 \phi_{ki}.$$

Note that $\chi(|\mathbf{x}_{ij}|) = 0$ if $|\mathbf{x}_{ij}| > 2r_0$. So

$$\mathcal{R}_2 \lesssim \frac{\varepsilon}{N^2} \sum_{i,j=1}^{N} |\mathbf{v}_{ij}|^4 \mathbb{1}_{|\mathbf{x}_{ij}| \le 2r_0} + \frac{1}{\varepsilon N^2} \sum_{i,k=1}^{N} |\mathbf{v}_{ki}|^2 \phi_{ki}$$

$$\le \frac{\varepsilon}{N^2} \sum_{i,j=1}^{N} |\mathbf{v}_{ij}|^4 \mathbb{1}_{|\mathbf{x}_{ij}| \le r_0} + \frac{\varepsilon}{N^2} \sum_{i,j=1}^{N} |\mathbf{v}_{ij}|^4 \mathbb{1}_{r_0 < |\mathbf{x}_{ij}| < 2r_0}$$

$$+ \frac{1}{\varepsilon N^2} \sum_{i,k=1}^{N} |\mathbf{v}_{ki}|^2 \phi_{ki}.$$

For ε sufficiently small, the first sum gets absorbed into the gain term (2.47). The second and third sums are dominated by $\mathcal{I}_2$. The term $\mathcal{R}_1$ can be estimated in exactly the same manner. We obtain

$$\frac{d}{dt}\mathcal{G}_3 \leq -c\Phi(\mathcal{D})\mathcal{V}_4 + a\mathcal{I}_2. \tag{2.48}$$

By analogy with the previous proof, we define the Lyapunov functional $\mathcal{L} = \mathcal{G}_3 + a\mathcal{V}_2$ and conclude that

$$\int_0^\infty \Phi(\mathcal{D}(t))\,\mathcal{V}_4(t)\,dt < \infty, \tag{2.49}$$

with the bound being independent of N. Furthermore, thanks to the uniform bound on the velocity, $\mathcal{D}(t) \leq ct + \mathcal{D}_0$. Thus,

$$\int_0^\infty \Phi(ct + \mathcal{D}_0)\mathcal{V}_4(t)\,dt < \infty. \tag{2.50}$$

By virtue of the heavy tail condition on Φ, $\mathcal{V}_4$ cannot be bounded away from the zero. And in view of its monotonicity, $\mathcal{V}_4 \to 0$. □

Remark 2.2 In each of the results above, we can actually extract a specific rate of alignment if we make an explicit assumption about the tail of the kernel. Thus, if

$$\phi(r) \sim \frac{1}{r^\beta}, \quad \forall r > r_0,$$

and $\beta \leq 1$, then from (2.49),

$$\int_1^\infty \frac{1}{t^\beta}\mathcal{V}_4(t)\,dt < \infty,$$

where $\mathcal{V}$ is the corresponding functional. So if $\beta = 1$, then there exists an $A > 0$ such that for any $T > 1$ there exists a $t \in [T, T^A]$ such that $\mathcal{V}_4(t) < \frac{1}{\ln t}$. Since $\ln t$ is proportional to $\ln T$ for all $t \in [T, T^A]$, this proves the above bound for all large times. If $\beta < 1$, then we argue that for some large $A > 0$ and all $T > 0$, we find $t \in [T, AT]$ such that $\mathcal{V}_4(t) \leq \frac{1}{t^{1-\beta}}$. But the latter is comparable for all values of $t \in [T, AT]$. Thus, we obtain the power rate as above for all t.

Let us summarize the obtained results:

$$\mathcal{V}_4(t) \lesssim \frac{1}{\ln t}, \quad \beta = 1,$$
$$\mathcal{V}_4(t) \lesssim \frac{1}{t^{1-\beta}}, \quad \beta < 1. \tag{2.51}$$

Of course the same argument applies to $\mathcal{V}_2$ under the conditions of part (ii).

2.8   Multi-Flocks, Clusters, and Multispecies

In the case when a flock is geometrically nonhomogeneous, the classical Cucker-Smale system may yield a nonrealistic dynamics. Let us consider one such example. Suppose all the masses are the same $m_i = \frac{1}{N}$, and suppose that we have two well-separated clusters in the flock, one containing a large collection of agents N_1, and one is small $N_2 \ll N_1$; see Fig. 2.4. Since masses represent influence strengths of the agents, one would expect that dynamics of the small flock will be largely independent of the large flock, and so its agent masses should rather be $\frac{1}{N_2}$. This is, however, contrary to what CS system would show: for any i in the small flock, since $\phi(\mathbf{x}_i - \mathbf{x}_j) \sim 0$ for all j in the large flock, we have

$$\dot{\mathbf{v}}_i \sim \frac{1}{N_1 + N_2} \sum_{j=1}^{N_1} \phi(\mathbf{x}_i - \mathbf{x}_j)(\mathbf{v}_j - \mathbf{v}_i) \sim 0.$$

So the dynamics of the small flock is stalled according to the classical CS description. To remedy the situation, Motsch and Tadmor in [75] proposed to consider a model where uniform averaging as in CS is replaced with adaptive averaging:

$$\dot{\mathbf{v}}_i = \frac{1}{\sum_k \phi(\mathbf{x}_i - \mathbf{x}_k)} \sum_{k=1}^{N} \phi(\mathbf{x}_i - \mathbf{x}_k)(\mathbf{v}_k - \mathbf{v}_i).$$

When applied to the described situation, we will have for each i in the small flock

$$\sum_k \phi(\mathbf{x}_i - \mathbf{x}_k) \sim N_2.$$

As a result, the model rebalances masses to the correct value $\frac{1}{N_2}$. More generally, one can consider the model with nonhomogeneous mass distribution:

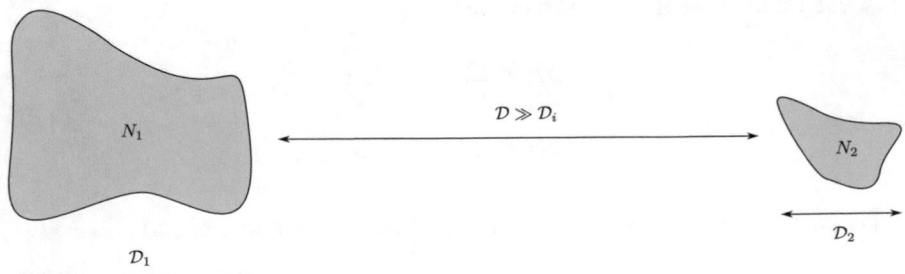

Fig. 2.4 Large flock overtakes dynamics of a small flock

$$\dot{\mathbf{v}}_i = \frac{\lambda}{\sum_k m_k \phi(\mathbf{x}_i - \mathbf{x}_k)} \sum_{k=1}^{N} m_k \phi(\mathbf{x}_i - \mathbf{x}_k)(\mathbf{v}_k - \mathbf{v}_i). \tag{2.52}$$

Despite the fact that the kernel is no longer symmetric in the Motsch-Tadmor model, one can still obtain the full analogue of Theorem 2.2:

$$\int_{\mathcal{D}_0}^{\infty} \phi(r)\, dr > \frac{\mathcal{A}_0 \|\phi\|_\infty}{\lambda} \quad \Rightarrow \quad \mathcal{A}(t) \le \mathcal{A}_0 e^{-t\lambda \|\phi\|_\infty^{-1} \phi(\overline{\mathcal{D}})}. \tag{2.53}$$

Notice that the mass is no longer present in the formula for the rate of convergence since it is simply scaled out of the new communication protocol. Indeed, arguing as before, we obtain

$$\frac{d}{dt} \ell(\mathbf{v}_{ij}) = \lambda \sum_{k=1}^{N} \frac{m_k \phi_{ik}}{\sum_{p=1}^{N} m_p \phi_{ip}} \ell(\mathbf{v}_{ki}) + \frac{m_k \phi_{jk}}{\sum_{p=1}^{N} m_p \phi_{jp}} \ell(\mathbf{v}_{jk})$$

$$\le \frac{\lambda}{M \|\phi\|_\infty} \phi(\mathcal{D}) \sum_{k=1}^{N} m_k [\ell(\mathbf{v}_{ki}) + \ell(\mathbf{v}_{jk})] = -\frac{\lambda}{\|\phi\|_\infty} \phi(\mathcal{D}) \ell(\mathbf{v}_{ij}).$$

The rest of the proof follows that of Theorem 2.2.

Due to lack of symmetry, the limiting velocity of the flock is no longer determined by the initial condition but rather becomes an *emergent* quantity of the dynamics. Since all the differences $\mathbf{v}_{ij}$ vanish exponentially fast, it implies that velocities do in fact converge to a time-independent limit as seen from integrating the velocity equation:

$$\lim_{t \to \infty} \mathbf{v}_i(t) = \mathbf{v}_i(0) + \lambda \int_0^\infty \sum_{k=1}^{N} \frac{m_k \phi_{ik}}{\sum_{p=1}^{N} m_p \phi_{ip}} \mathbf{v}_{ki}(s)\, ds.$$

Let us recognize still one more issue of the alignment formula (2.53)—the rate depends on the global bound on the diameter of the full system $\overline{\mathcal{D}}$. While global alignment may as well be slow for separated flocks, the alignment within each flock may happen at a much faster rate proportional to $\phi(\mathcal{D}_i)$. In order to capture this phenomenon, we need to incorporate multi-scale communication into the model. Let us assume that we have A well-separated flocks indexed by $\alpha = 1, \ldots, A$. We have velocities $\mathbf{v}_{\alpha i}$ and positions $\mathbf{x}_{\alpha i}$ within each flock indexed by $i = 1, \ldots, N_\alpha$. When $\alpha \ne \beta$, the communication between flocks can be approximated by

$$\phi(\mathbf{x}_{\alpha i} - \mathbf{x}_{\beta j}) \sim \phi(\mathbf{X}_\alpha - \mathbf{X}_\beta),$$

where $\mathbf{X}_\alpha$ is the center of mass, $\mathbf{X}_\alpha = \frac{1}{M_\alpha} \sum_{i=1}^{N_\alpha} m_{\alpha i} \mathbf{x}_{\alpha i}$. Consequently,

$$\dot{\mathbf{v}}_{\alpha i} = \lambda \sum_{j=1}^{N_\alpha} m_{\alpha j} \phi(\mathbf{x}_{\alpha i} - \mathbf{x}_{\alpha j})(\mathbf{v}_{\alpha j} - \mathbf{v}_{\alpha i}) + \varepsilon \sum_{\beta \neq \alpha} M_\beta \phi(\mathbf{X}_\alpha - \mathbf{X}_\beta)(\mathbf{V}_\beta - \mathbf{v}_{\alpha i}),$$

where $\mathbf{V}_\alpha = \frac{1}{M_\alpha} \sum_{i=1}^{N_\alpha} m_{\alpha i} \mathbf{v}_{\alpha i}$, and $\varepsilon \ll \lambda$. To make the model more flexible, we may assume that communications within flocks may be different and also distinct from inter-flock communication. We thus consider the following model:

$$\begin{cases} \dot{\mathbf{x}}_{\alpha i} = \mathbf{v}_{\alpha i}, \\[2mm] \dot{\mathbf{v}}_{\alpha i} = \lambda_\alpha \sum_{j=1}^{N_\alpha} m_{\alpha j} \phi_\alpha(\mathbf{x}_{\alpha i} - \mathbf{x}_{\alpha j})(\mathbf{v}_{\alpha j} - \mathbf{v}_{\alpha i}) \\[4mm] \qquad\quad + \varepsilon \sum_{\beta \neq \alpha} M_\beta \Psi(\mathbf{X}_\alpha - \mathbf{X}_\beta)(\mathbf{V}_\beta - \mathbf{v}_{\alpha i}). \end{cases} \tag{2.54}$$

It follows from the construction that the macroscopic variables $\mathbf{X}_\alpha$, $\mathbf{V}_\alpha$ satisfy the upscaled system:

$$\begin{cases} \dot{\mathbf{X}}_\alpha = \mathbf{V}_\alpha, \\[2mm] \dot{\mathbf{V}}_\alpha = \varepsilon \sum_{\beta \neq \alpha} M_\beta \Psi(\mathbf{X}_\alpha - \mathbf{X}_\beta)(\mathbf{V}_\beta - \mathbf{V}_\alpha). \end{cases} \tag{2.55}$$

To study the collective behavior of the multi-flock system (2.54), let us consider the following size parameters:

$$\mathcal{D}_\alpha = \max_{i,j} |\mathbf{x}_{\alpha i} - \mathbf{x}_{\alpha j}|, \quad \mathcal{D} = \max_{\alpha, \beta} |\mathbf{X}_\alpha - \mathbf{X}_\beta|,$$

$$\mathcal{A}_\alpha = \max_{i,j} |\mathbf{v}_{\alpha i} - \mathbf{v}_{\alpha j}|, \quad \mathcal{A} = \max_{\alpha, \beta} |\mathbf{V}_\alpha - \mathbf{V}_\beta|.$$

The alignment of macroscopic quantities follows from the same system of ODIs as we derived in the classical case:

$$\begin{cases} \dot{\mathcal{A}} \leq -\varepsilon M \Psi(\mathcal{D})\mathcal{A}, \\[2mm] \dot{\mathcal{D}} \leq \mathcal{A}. \end{cases} \tag{2.56}$$

Thus, under heavy tail condition on Ψ, the consensus directions $\mathbf{V}_\alpha$ will in fact align exponentially fast according to Theorem 2.2. To understand the alignment within each flock, we consider a maximizing triple ℓ, i, j for an α-flock and apply computation (2.23):

$$\frac{d}{dt}\ell(\mathbf{v}_{\alpha i} - \mathbf{v}_{\alpha j}) \leq -\lambda_\alpha M_\alpha \phi(\mathcal{D}_\alpha)\mathcal{A}_\alpha - \varepsilon M \Psi(\mathcal{D})\mathcal{A}_\alpha.$$

Consequently, we obtain the following system of ODIs:

$$
\begin{cases}
\dot{\mathcal{A}}_\alpha \leq -\lambda_\alpha M_\alpha \phi(\mathcal{D}_\alpha)\mathcal{A}_\alpha - \varepsilon M \Psi(\mathcal{D})\mathcal{A}_\alpha, \\
\dot{\mathcal{D}}_\alpha \leq \mathcal{A}_\alpha, \\
\dot{\mathcal{A}} \leq -\varepsilon M \Psi(\mathcal{D})\mathcal{A}, \\
\dot{\mathcal{D}} \leq \mathcal{A}.
\end{cases}
\tag{2.57}
$$

Ignoring $-\varepsilon M \Psi(\mathcal{D})\mathcal{A}_\alpha$ in the $\mathcal{A}_\alpha$ equation for the moment, we see that the α-flock decouples from the rest. Consequently, a fast internal alignment ensues by virtue of Theorem 2.2.

Theorem 2.7 (Fast Internal Flocking) *If for a given $\alpha \in \{1, \ldots, A\}$ the kernel ϕ_α has a heavy tail, the α-flock aligns exponentially fast at a rate functionally dependent on ϕ_α, initial data, and λ_α:*

$$
\max_i |\mathbf{v}_{\alpha i}(t) - \mathbf{V}_\alpha(t)| \lesssim e^{-\delta t}.
$$

It is interesting to note that this alignment process is completely independent from the inter-flock communication. So long range internal communication leads to local emergence despite potentially destabilizing influence of the outside crowd. On the other hand, if the inter-flock communication Ψ is global, e.g., it satisfies the heavy tail condition (2.6), then the global alignment occurs even if internal communications are weak or even absent. This is clear from (2.57) if we drop $-\lambda_\alpha M_\alpha \phi(\mathcal{D}_\alpha)\mathcal{V}_\alpha$ and conclude boundedness of $\mathcal{D}$ from the last two equations. The alignment rate in this case is global but occurs on the slow time scale.

Theorem 2.8 (Slow Global Flocking) *Assuming that Ψ has a heavy tail and all $\phi_\alpha \geq 0$, all solutions to (2.54) align exponentially fast at a rate functionally dependent on Ψ, initial data, and ε:*

$$
\max_{\alpha, i} |\mathbf{v}_{\alpha i}(t) - \mathbf{V}| \lesssim e^{-\delta t}.
$$

Asymptotic dependence of the implied alignment rates for small ε and large λ_α for the Cucker-Smale kernel can be readily obtained from formulas (2.26) and (2.27). Thus, in the context of fast local alignment, we obtain $\delta \sim \lambda_\alpha$ for all $\beta \leq 1$, while in the context of slow alignment, $\delta \sim \varepsilon^{\frac{1}{1-\beta}}$ for $\beta < 1$, and $\delta \sim \varepsilon e^{-1/\varepsilon}$ for $\beta = 1$.

It is sometimes convenient to pass to the reference frame evolving with the momentum and center of mass of the flock:

$$
\mathbf{w}_{\alpha i} = \mathbf{v}_{\alpha i} - \mathbf{V}_\alpha, \quad \mathbf{y}_{\alpha i} = \mathbf{x}_{\alpha i} - \mathbf{X}_\alpha.
\tag{2.58}
$$

Using (2.54) and (2.55), one readily obtains the system:

$$\begin{cases} \dot{\mathbf{y}}_{\alpha i} = \mathbf{w}_{\alpha i}, \\ \\ \dot{\mathbf{w}}_{\alpha i} = \lambda_\alpha \sum_{j=1}^{N_\alpha} m_{\alpha j} \phi_{\alpha i j} (\mathbf{w}_{\alpha i} - \mathbf{w}_{\alpha j}) - \varepsilon R_\alpha(t) \mathbf{w}_{\alpha i}, \end{cases} \qquad (2.59)$$

where

$$R_\alpha(t) = \sum_{\beta \neq \alpha} M_\beta \Psi (\mathbf{X}_\alpha - \mathbf{X}_\beta). \qquad (2.60)$$

We used a shortcut to denote $\phi_{\alpha i j} = \phi_\alpha (\mathbf{y}_{\alpha i} - \mathbf{y}_{\alpha j})$. The following lemma is straightforward.

Lemma 2.3 *The old set of variables* $(\mathbf{x}_{\alpha i}, \mathbf{v}_{\alpha i})_{\alpha, i}$ *satisfies* (2.54) *if and only if the new set* $(\mathbf{y}_{\alpha i}, \mathbf{w}_{\alpha i})_{\alpha, i}$ *satisfies* (2.59) *and the macroscopic variables* $(\mathbf{X}_\alpha, \mathbf{V}_\alpha)_\alpha$ *satisfy* (2.55).

An immediate consequence of (2.59) is the maximum principle within each flock relative to its momentum. In particular, we obtain a class of solutions, called Mikado flocks, given by

$$\mathbf{w}_{\alpha i}(t) = w_{\alpha i}(t) \mathbf{r}_\alpha,$$

where $\mathbf{r}_\alpha$ are fixed unit vectors. We will address those in more detail in the context of hydrodynamic systems in Sect. 9.2.

Another type of cluster-adapted communication appears in multispecies models. Here, each flock α represents a cluster of species where communication between agents inside each cluster is homogeneous as in classical CS and communication between species is regulated by a different set of kernels:

$$\dot{\mathbf{v}}_{\alpha i} = \sum_{\beta=1}^{A} \frac{1}{N_\beta} \sum_{j=1}^{N_\beta} \phi_{\alpha\beta} (\mathbf{x}_{\alpha i} - \mathbf{x}_{\beta j}) (\mathbf{v}_{\beta j} - \mathbf{v}_{\alpha i}). \qquad (2.61)$$

One can easily run the same scheme and prove alignment with a rather rough condition: the heavy tail of the kernel $\phi(r) = \min_{\alpha, \beta} \phi_{\alpha\beta}(r)$. A more subtle spectral condition can be obtained in terms of the weighted graph Laplacian associated with the matrix kernel $\Phi = (\phi_{\alpha\beta})$:

$$(\Delta\Phi(r))_{\alpha\beta} = \begin{cases} -\phi_{\alpha\beta}(r)\sqrt{M_\alpha M_\beta}, & \alpha \neq \beta, \\ \\ \sum_{\gamma \neq \alpha} M_\gamma \phi_{\alpha\gamma}(r), & \alpha = \beta. \end{cases} \qquad (2.62)$$

Namely, the condition requires the second eigenvalue of $\Delta\Phi(r)$, denoted $\lambda_2(\Delta\Phi(r))$, to have heavy tail:

$$\lambda_2(\Delta\Phi(r)) \gtrsim \frac{1}{(1+r)^{1-\delta}}.$$

Then the energy of fluctuations of each flock converges to zero exponentially fast: $\mathcal{V}_2(\alpha) \lesssim e^{-\delta t}$. We refer the reader to [53] for details.

2.9 Notes and References

System (2.1) with smooth kernel (1.3) was first introduced by Cucker and Smale in [33, 34] where Theorem 2.1 was established using a graph theoretical approach. This was one of the first models which featured unconditional alignment, which is partly responsible for its success. Detailed surveys are given in [1, 73, 76, 102]. Shortly after, an analytical proof of the more general Theorem 2.2 was found by Ha and Liu, [43], which has undergone several iterations in the literature. Stability estimates appeared in Ha et al. [49]. We will return to them repeatedly in the kinetic and hydrodynamic contexts.

Collision avoidance in the context of CS systems was first addressed in Park et al. [80] with the use of an inter-particle bonding force showing no collision for two agent configurations. Cucker and Dong [31] proposed a simpler model with unconditional avoidance incorporating a singular repulsion force; see Chap. 3. In 1D, the criterion of Theorem 2.4 appeared in [50]. For unforced systems, the use of singular communication was proposed by Carrillo, Choi, Mucha, and Peszek [16]. The result has been extended to models with nonlinear couplings by Markou, [72] and to various forced systems by Peszek et al. [26, 61]. For weakly singular kernels $\beta < 1$ despite the presence of collisions, Peszek developed a theory of weak solutions to system (2.1) [82, 83]. The estimates (2.37)–(2.39) are slight improvements over [16] suitable for applications to the corrector method, which was introduced in Dietert and Shvydkoy [38] as a way to handle degenerate coercivity. For local kernels, flocking dynamics can also be achieved by steering the system with a centralized control; see Capnigro et al. [12]. Moreover, any given collective outcome can also be achieved using decentralized control algorithm [26].

Motsch-Tadmor model was first introduced in [75] and subsequently appeared in several applications; see [76]. It plays a further role in passing to the hydrodynamic limit in multi-scale models to recover the pressureless Euler alignment system, which we discuss in detail in later chapters. Multi-flocks are introduced in [96] with Mikado solutions presented in [65]. Multispecies models appeared earlier in [53].

Chapter 3
Forced Systems

The Cucker-Smale alignment force is typically supplemented by other forces in more realistic complex models. The theoretical studies of animal behavior distinguish three basic regions of mutual interactions between agents. The first inner region is dominated by repulsion forces, driven by the fear of collisions. The second intermediate region is characterized by alignment relative to the direction of motion of other objects, which is where Cucker-Smale force plays the dominant role. And the third outer region is characterized by tendencies to attract and stay within the group. This 3-zone model of interactions has been fundamental in many studies in behavioral biology [3, 20, 21, 23, 30, 54]. A notable implementation of the 3-zone model to computer animation of bird flight was proposed by C. W. Reynolds in [87], which was subsequently applied in motion picture production (Fig. 3.1).

More refined modeling such as inclusion of cone of vision into the alignment zone, nearest neighbor rule, self-propulsion, friction, noise, and external confinement allow to replicate some of the interesting behavioral phenomena such as milling patterns and aggregation. A model with a wealth of such outcomes was proposed by D'Orsogna et al. [27]:

$$
\begin{cases}
\dot{\mathbf{x}}_i = \mathbf{v}_i, \\[2mm]
\dot{\mathbf{v}}_i = (\alpha - \beta |\mathbf{v}_i|^2)\mathbf{v}_i - \dfrac{1}{N}\sum_{j=1}^{N} \nabla U(\mathbf{x}_i - \mathbf{x}_j).
\end{cases}
$$

We refer to [14, 17] for comprehensive surveys of models with attraction/repulsion forces.

The purpose of this chapter is to study the effects of these additional forces in the context of alignment models and see how flocking dynamics changes or persists. The center of our attention will be attraction/repulsion potential forces and the method of hypocoercivity which allows to produce N-independent estimates suitable for large crowd limit $N \to \infty$. The most well-understood case here is the 2-

© Springer Nature Switzerland AG 2021

R. Shvydkoy, *Dynamics and Analysis of Alignment Models of Collective Behavior*,
Nečas Center Series, https://doi.org/10.1007/978-3-030-68147-0_3

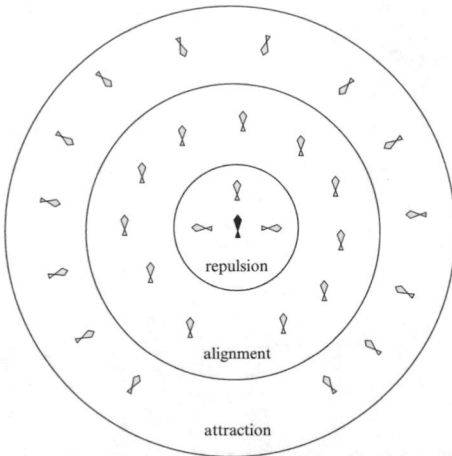

Fig. 3.1 3-Zone model

zone attraction/alignment model. In the presence of repulsion or full 3-zone model, only partial N-dependent results are known. Much of the difficulty of the 3-zone model boils down to nonexistence of naturally aligned steady states that minimize total energy. Say, if $U \geq 0$ is a radial potential with $U = 0$ in the alignment annulus $|\mathbf{x}| \in [r_0, r_1]$, then all separations of such stationary solutions will need to fit within it, $|\mathbf{x}_{ij}| \in [r_0, r_1]$. This is geometrically impossible unless $N \leq N(r_0, r_1, n)$. For this reason, we do not expect to have any strong flocking behavior. The repulsion itself acts as a mechanism of expansion of the flock which also presents an obstacle in obtaining N-dependent bounds on the diameter. So we will proceed being mindful of these mathematical limitations and start with establishing general flocking results for 3-zone models.

3.1   3-Zone Model: Small Crowd Flocking

Let us consider the Cucker-Smale system with a general potential interaction force:

$$\begin{cases} \dot{\mathbf{x}}_i = \mathbf{v}_i, \\[2mm] \dot{\mathbf{v}}_i = \dfrac{1}{N} \sum_{j=1}^{N} \phi(\mathbf{x}_i - \mathbf{x}_j)(\mathbf{v}_j - \mathbf{v}_i) - \dfrac{1}{N} \sum_{j=1}^{N} \nabla U(\mathbf{x}_i - \mathbf{x}_j). \end{cases} \tag{3.1}$$

Here, we can work with a rather general radially symmetric potential $U \in C^1$ as long as it forms a singularity at the origin and at infinity:

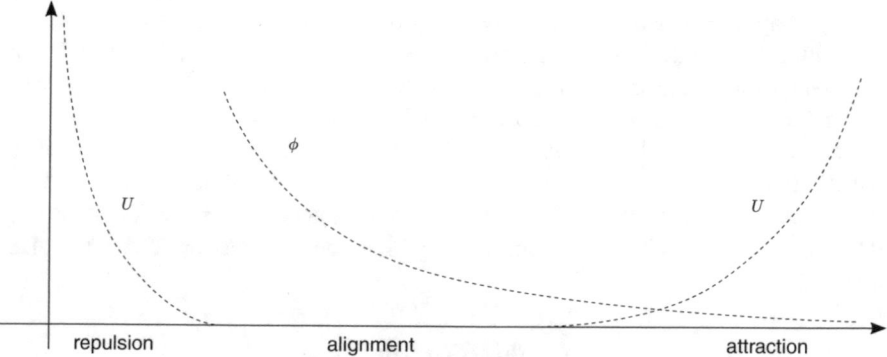

Fig. 3.2 Alignment versus potential in the 3-zone model

$$\lim_{r \to 0, \infty} U(r) = \infty. \tag{3.2}$$

We also assume that

$$\phi \in C^1, \quad \phi(r) > 0 \text{ for all } r > r_0, \quad \lim_{r \to \infty} \phi(r) = 0. \tag{3.3}$$

Although necessarily there is an overlap of alignment and attraction ranges, we do not assume any specific rate. So the kernel can decay as fast as possible as long as it remains positive in the long range. In other words, alignment action is mostly confined to an intermediate region which is consistent with its placement in the 3-zone model. Figure 3.2 illustrates a typical picture we have in mind.

System (3.1) is Galilean invariant and conserves momentum. So as always we can assume $\bar{\mathbf{v}} = 0$. The total energy is now composed of kinetic and potential parts:

$$\mathcal{E} = \mathcal{K} + \mathcal{P},$$

$$\mathcal{K} = \frac{1}{2N} \sum_{i=1}^{N} |\mathbf{v}_i|^2 = \frac{1}{4N^2} \sum_{i=1}^{N} |\mathbf{v}_{ij}|^2, \quad \mathcal{P} = \frac{1}{N^2} \sum_{i,j=1}^{N} U(\mathbf{x}_{ij}). \tag{3.4}$$

We have the energy law:

$$\frac{\mathrm{d}}{\mathrm{d}t} \mathcal{E} = -\mathcal{I}_2. \tag{3.5}$$

In particular, it implies the $\mathcal{P}$ remains bounded, which, in view of (3.2), implies that that flock remains bounded, $\mathcal{D}(t) \le \overline{\mathcal{D}}_N$, and non-collisional. Note that the upper bound depends on N, so such conclusions would not hold in the large crowd limit $N \to \infty$. That is why we consider them more suitable for small crowd applications.

The energy law above is lacking coercivity for two reasons—it is missing the potential energy entirely on the right-hand side, and the communication bound

$\phi(\overline{\mathcal{D}}_N)$ may in fact vanish due to degeneracy of the kernel. In the next two sections, we will demonstrate how the potential energy can be recovered in the 2-zone case when only attraction is present. In general, we can only fill in the missing communication using our corrector method developed in Sect. 2.7.

Indeed, let us recall the estimate (2.48). In its derivation, we have not used the equation except the maximum principle. In view of the global bound on K, we still have a bound on each velocity $|\mathbf{v}_i| < C_N$ by a constant depending on N. So the only difference for us will be in the constant a, which now depends on N. We therefore arrive at

$$\int_0^\infty \Phi(\mathcal{D})\mathcal{V}_4(t)\,\mathrm{d}t < \infty.$$

Given that $\Phi > 0$ and $\mathcal{D}$ is bounded, we conclude that

$$\int_0^\infty \mathcal{V}_4(t)\,\mathrm{d}t < \infty.$$

It is also straightforward to see that $\left|\dfrac{\mathrm{d}}{\mathrm{d}t}\mathcal{V}_4\right| \leq C_N$. So $\mathcal{V}_4$ is in fact a uniformly continuous function. Together with integrability, it implies that $\mathcal{V}_4 \to 0$, and hence, the system aligns.

As noted in the introduction, it is unlikely that substantially faster convergence can be proved in general, as the system is not likely to flock strongly.

Theorem 3.1 *Any 3-zone model* (3.1) *with potential satisfying* (3.2) *and kernel satisfying* (3.3) *aligns* $\mathcal{A}(t) \to 0$ *and flocks* $\mathcal{D}(t) \leq \overline{\mathcal{D}}_N$, *although with rates and bounds which may depend on* N.

Even if the kernel ϕ was non-degenerate, we could have not improved the result. The natural way would be to operate with the energy law

$$\frac{\mathrm{d}}{\mathrm{d}t}\mathcal{E} \leq -c_0\mathcal{K}$$

and conclude that $\int_0^\infty \mathcal{K}(t)\,\mathrm{d}t < \infty$. This is still insufficient for strong flocking since the residual velocities would stir the system ever so slightly. The force can be modified to avoid this issue by having an additional kinetic factor:

$$\mathbf{F}_i = -\sqrt{\mathcal{V}_2}\frac{1}{N}\sum_{j=1}^N \nabla U(\mathbf{x}_i - \mathbf{x}_j). \tag{3.6}$$

This model was considered by Cucker and Dong [31] as a way to obtain strong flocking and collision avoidance in repulsion-only setting, $\mathrm{Supp}\,U \subset \{r < r_0\}$. One can argue that it is in fact natural for repulsive forces to subside for still crowd since there is no reason for agents to panic if they are not on a collision course. The same perhaps cannot be said for the attraction force since the tendency to stay together would induce motion even in a still but dispersed flock. Nonetheless, let us discuss this kinetic forcing in general.

Here, we assume that communication is non-degenerate and either $U \to \infty$ as $r \to \infty$ or ϕ has a monotone heavy tail. The energy law reads

$$\frac{d}{dt}(\sqrt{\mathcal{K}} + \mathcal{P}) = -\frac{\mathcal{I}}{\sqrt{\mathcal{K}}}.$$

So in the first case, we conclude boundedness of the diameter as earlier. In the second, we obtain

$$\int_0^\infty \phi(\mathcal{D}(t))\sqrt{\mathcal{K}(t)}\, dt < \infty.$$

Noting that $\dfrac{d}{dt}\mathcal{D} \leq C_N\sqrt{\mathcal{K}(t)}$, we further obtain

$$\int_{\mathcal{D}_0}^{\mathcal{D}(t)} \phi(r)\, dr < \infty.$$

This immediately implies $\mathcal{D}(t) \leq \overline{\mathcal{D}}_N$. Bounded diameter implies an integrability condition on the kinetic energy:

$$\int_0^\infty \sqrt{\mathcal{K}(t)}\, dt < \infty.$$

In particular, this implies that $\mathcal{A}$ is integrable and as a consequence, strong flocking ensues. The convergence $\mathcal{A} \to 0$ follows as before from uniform continuity.

Theorem 3.2 *Consider system* (3.1) *with non-degenerate $\phi > 0$ and either a strong potential $U(r) \to \infty$, $r \to \infty$, or monotonically decaying heavy tail ϕ. Then any solution aligns $\mathcal{A} \to 0$ and flocks strongly, $\mathbf{x}_i(t) - \mathbf{x}_j(t) \to \bar{\mathbf{x}}_{ij}$.*

Let us note that the geometric obstruction to flocking we alluded to earlier does not apply in this case since the system has arbitrarily dispersed fully aligned steady states.

3.2   External Confinement: Hypocoercivity

In this and next sections, we will consider 2-zone models where non-degenerate alignment is combined with potential attraction forces. The goal will be to show that such models exhibit much more robust flocking properties with the results being N-independent and extendable to macroscopic models as well.

Let us first illustrate the method on a simpler example of a confinement force:

$$\begin{cases} \dot{\mathbf{x}}_i = \mathbf{v}_i, \\[2mm] \dot{\mathbf{v}}_i = \dfrac{1}{N} \sum_{j=1}^{N} \phi(\mathbf{x}_i - \mathbf{x}_j)(\mathbf{v}_j - \mathbf{v}_i) - \nabla U(\mathbf{x}_i), \end{cases} \tag{3.7}$$

where U is a convex radially increasing potential. While in general one can perform analysis in the case of a strictly convex potential, for the sake of brevity, we focus on one particular case, that of a quadratic U:

$$U(\mathbf{x}) = \frac{1}{2}|\mathbf{x}|^2.$$

In this case, the system reads

$$\begin{cases} \dot{\mathbf{x}}_i = \mathbf{v}_i, \\[2mm] \dot{\mathbf{v}}_i = \dfrac{1}{N} \sum_{j=1}^{N} \phi(\mathbf{x}_i - \mathbf{x}_j)(\mathbf{v}_j - \mathbf{v}_i) - \mathbf{x}_i. \end{cases} \tag{3.8}$$

The natural limiting state of this system is not, in fact, alignment but rather a harmonic oscillator. Indeed, denoting as before the center of mass and momentum by $\bar{\mathbf{x}}, \bar{\mathbf{v}}$, respectively, we obtain

$$\frac{\mathrm{d}}{\mathrm{d}t}\bar{\mathbf{x}} = \bar{\mathbf{v}}, \quad \frac{\mathrm{d}}{\mathrm{d}t}\bar{\mathbf{v}} = -\bar{\mathbf{x}}. \tag{3.9}$$

Also due to the linear nature of the forcing, we can shift the system of coordinates to $(\bar{\mathbf{x}}, \bar{\mathbf{v}})$ and assume that $\bar{\mathbf{x}} = 0, \bar{\mathbf{v}} = 0$. In this case, system (3.8) is technically equivalent to the system with interaction forces (3.1) with the same quadratic potential. So the question becomes to show that the solution tends to zero in the new reference frame (still denoted $(\mathbf{x}, \mathbf{v})$).

The full energy of the system is now given by

$$\mathcal{E} = \mathcal{K} + \mathcal{P},$$

$$\mathcal{K} = \frac{1}{2N} \sum_{i=1}^{N} |\mathbf{v}_i|^2, \quad \mathcal{P} = \frac{1}{2N} \sum_{i=1}^{N} |\mathbf{x}_i|^2.$$

It will also be important to consider the "particle energy", i.e., the L^∞-version of the energy:

$$\mathcal{E}_\infty = \frac{1}{2} \max_i (|\mathbf{v}_i|^2 + |\mathbf{x}_i|^2). \tag{3.10}$$

Theorem 3.3 *Suppose that the kernel ϕ is bounded and decreasing and satisfies the weak heavy tail condition:*

$$\int_0^\infty r\phi(r)\,dr = \infty. \tag{3.11}$$

Then system (3.8) settles on its harmonic oscillator (3.9) exponentially fast, meaning

$$\max_i(|\mathbf{v}_i(t) - \bar{\mathbf{v}}(t)|^2 + |\mathbf{x}_i(t) - \bar{\mathbf{x}}(t)|^2) \le Ce^{-\delta t},$$

for some $\delta > 0$ and $C = C(\mathbf{v}_0, \mathbf{x}_0, \phi)$ independent of N.

Proof The result amounts to establishing an exponential bound on $\mathcal{E}_\infty$.
Straight from the equations, we obtain the energy law:

$$\frac{d}{dt}\mathcal{E} = -\mathcal{I}_2 \le -\phi(\mathcal{D})\mathcal{K}. \tag{3.12}$$

Note that the dissipation is not coercive even if we knew a lower bound $\phi(\mathcal{D}) > c_0$. However, we proceed with $\mathcal{E}_\infty$:

$$\frac{d}{dt}\mathcal{E}_\infty \le \frac{1}{N}\sum_j \phi_{ij}(\mathbf{v}_j - \mathbf{v}_i)\cdot\mathbf{v}_i \le \frac{1}{2N}\sum_j \phi_{ij}(|\mathbf{v}_j|^2 - |\mathbf{v}_i|^2)$$

$$\le \frac{1}{2N}\sum_j \phi_{ij}|\mathbf{v}_j|^2 \le \|\phi\|_\infty\mathcal{K}.$$

To combine the two equations into one system, let us note that $\mathcal{D} \le 4\sqrt{\mathcal{E}_\infty}$. Thus,

$$\frac{d}{dt}\mathcal{E} \le -\phi(4\sqrt{\mathcal{E}_\infty})\mathcal{K}$$

$$\frac{d}{dt}\mathcal{E}_\infty \le C\mathcal{K}.$$

Consider the Lyapunov function:

$$\mathcal{L} = \mathcal{E} + \frac{1}{C}\int_0^{\mathcal{E}_\infty} \phi(4\sqrt{r})\,dr.$$

From our heavy tail assumption, it follows that $\mathcal{E}_\infty$ remains bounded, from which we conclude that $\mathcal{D}(t) \le \bar{\mathcal{D}}$. Now the energy law (3.12) reads

$$\frac{d}{dt}\mathcal{E} \le -c_0\mathcal{K}. \tag{3.13}$$

Next, we proceed with a hypocoercivity argument to restore $\mathcal{K}$ on the right-hand side to the full energy $\mathcal{E}$. We consider the corrector (longitudinal momentum):

$$\mathcal{X} = \frac{1}{N} \sum_{i=1}^{N} \mathbf{x}_i \cdot \mathbf{v}_i,$$

and note that for $\lambda < \frac{1}{4}$, $\mathcal{E} + \lambda \mathcal{X} \sim \mathcal{E}$. Let us compute the derivative:

$$\frac{d}{dt}\mathcal{X} = \frac{1}{N^2} \sum_{j,i} \phi_{ij}(\mathbf{v}_j - \mathbf{v}_i) \cdot \mathbf{x}_i + \frac{1}{N} \sum_i (|\mathbf{v}_i|^2 - |\mathbf{x}_i|^2)$$

$$\leq \frac{4\|\phi\|_\infty^2}{N^2} \sum_{j,i} (|\mathbf{v}_j|^2 + |\mathbf{v}_i|^2) + \frac{1}{2N} \sum_{j,i} |\mathbf{x}_i|^2 + \frac{1}{N} \sum_i (|\mathbf{v}_i|^2 - |\mathbf{x}_i|^2)$$

$$\leq C\mathcal{K} - \mathcal{P}.$$

Choosing $\lambda < c_0/2C$, we obtain

$$\frac{d}{dt}(\mathcal{E} + \lambda \mathcal{X}) \leq -c_1 \mathcal{E} \sim -(\mathcal{E} + \lambda \mathcal{X}).$$

This establishes exponential decay of $\mathcal{E}$.

Now that the L^2-energy is decaying exponentially, we can extrapolate to obtain exponential decay for $\mathcal{E}_\infty$ as well. The method is similar—we consider an amended version of $\mathcal{E}_\infty$:

$$\mathcal{E}_\infty^\lambda = \frac{1}{2} \max_i (|\mathbf{v}_i|^2 + |\mathbf{x}_i|^2 + \lambda \mathbf{x}_i \cdot \mathbf{v}_i).$$

Again, for $\lambda < \frac{1}{4}$, this does not alter the particle energy much, $\mathcal{E}_\infty \sim \mathcal{E}_\infty^\lambda$. Differentiating at a point of maximum, we obtain

$$\frac{d}{dt}\mathcal{E}_\infty^\lambda = \frac{1}{N} \sum_j \phi_{ij}(\mathbf{v}_j - \mathbf{v}_i) \cdot (\mathbf{v}_i + \lambda \mathbf{x}_i) + \lambda |\mathbf{v}_i|^2 - \lambda |\mathbf{x}_i|^2$$

$$\leq \frac{1}{N} \sum_j \phi_{ij} \mathbf{v}_j \cdot \mathbf{v}_i - \frac{1}{N} \sum_j \phi_{ij} |\mathbf{v}_i|^2 + \lambda \frac{1}{N} \sum_j \phi_{ij} \mathbf{v}_j \cdot \mathbf{x}_i$$

$$- \lambda \frac{1}{N} \sum_j \phi_{ij} \mathbf{v}_i \cdot \mathbf{x}_i + \lambda |\mathbf{v}_i|^2 - \lambda |\mathbf{x}_i|^2.$$

In the gain term $-\frac{1}{N} \sum_j \phi_{ij} |\mathbf{v}_i|^2$, we replace ϕ_{ij} by a lower bound c_0:

$$-\frac{1}{N}\sum_j \phi_{ij}|\mathbf{v}_i|^2 \le -c_0|\mathbf{v}_i|^2,$$

and in the rest, we simply use boundedness of the kernel. Hence, if λ is small enough, the term $\lambda|\mathbf{v}_i|^2$ gets absorbed. We obtain at this point

$$\frac{d}{dt}\mathcal{E}_\infty^\lambda \lesssim \frac{1}{N}\sum_j |\mathbf{v}_j||\mathbf{v}_i| + \lambda\frac{1}{N}\sum_j |\mathbf{v}_j||\mathbf{x}_i| + \lambda\frac{1}{N}\sum_j |\mathbf{v}_i||\mathbf{x}_i| - c_1|\mathbf{v}_i|^2 - \lambda|\mathbf{x}_i|^2$$

$$\le 4\mathcal{K} + \frac{c_1}{4}|\mathbf{v}_i|^2 + 4\lambda\mathcal{K} + \frac{1}{4}\lambda|\mathbf{x}_i|^2 + \lambda|\mathbf{v}_i|^2 + \frac{1}{4}\lambda|\mathbf{x}_i|^2 - c_1|\mathbf{v}_i|^2 - \lambda|\mathbf{x}_i|^2$$

$$\lesssim \mathcal{K} - |\mathbf{v}_i|^2 - |\mathbf{x}_i|^2 \lesssim \mathcal{K} - \mathcal{E}_\infty^\lambda.$$

Since we already know that $\mathcal{K}$ is exponentially decaying, this establishes a similar bound on $\mathcal{E}_\infty^\lambda$. $\square$

Let us note that for a small number of agents, when the dependence on N is not an issue, one can actually obtain a much stronger result: as long as $\phi(r) > 0$ for all $r > 0$, the conclusion of the theorem holds true. Indeed, from the first lines, we have established that the total energy is decaying and, hence, bounded. But from the potential part $\mathcal{P}$, this immediately implies that the flock is bounded, with a bound depending on N. This sets the rest of the argument to go through.

This observation clearly shows that it is impossible to construct a simple example with a few agents, like we did in Example 2.1, to prove sharpness of the heavy tail condition (3.11).

3.3   2-Zone Model: Attraction + Alignment

We consider the system with pairwise interactions determined by a radially symmetric smooth potential $U \in C^2(\mathbb{R}_+)$ (Fig. 3.3):

$$\begin{cases} \dot{\mathbf{x}}_i = \mathbf{v}_i, \\ \dot{\mathbf{v}}_i = \frac{1}{N}\sum_{j=1}^N \phi(\mathbf{x}_i - \mathbf{x}_j)(\mathbf{v}_j - \mathbf{v}_i) - \frac{1}{N}\sum_{j=1}^N \nabla U(\mathbf{x}_i - \mathbf{x}_j). \end{cases} \tag{3.14}$$

Notice first that system (3.14) preserves momentum and is Galilean invariant. So we can shift the center of mass and momentum to zero. The energy still satisfies the classical law:

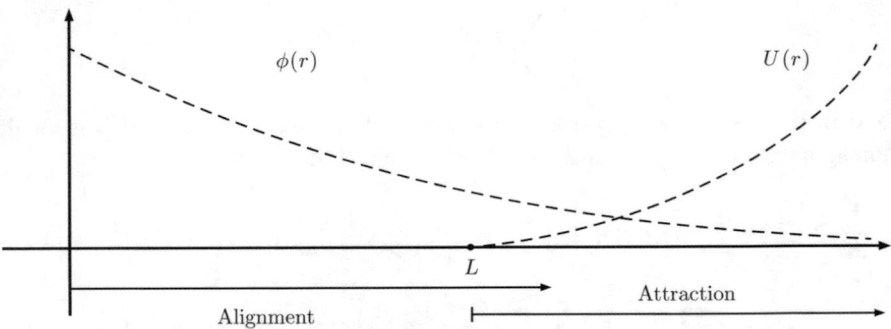

Fig. 3.3 2-Zone attraction-alignment model

$$\frac{\mathrm{d}}{\mathrm{d}t}\mathcal{E} = -\frac{1}{2N^2}\sum_{i,j=1}^{N}\phi_{ij}|\mathbf{v}_{ij}|^2 = -\mathcal{I}. \tag{3.15}$$

In this section, we prove an alignment with algebraic rate for the 2-zone attraction/alignment model with non-degenerate communication kernel. Let us recall the notation for Japanese brackets used throughout:

$$\langle r \rangle = (1 + r^2)^{\frac{1}{2}}.$$

Theorem 3.4 *Suppose the kernel and potential satisfy the following power-law assumptions:*

$$\phi'(r) \le 0, \quad \phi(r) \ge \frac{c_0}{\langle r \rangle^\gamma}, \ for \ r \ge 0,$$

and for some $\beta > 1$ and $L' > L > 0$,

> *Support:* $U \in C^2(\mathbb{R}^+), \quad U(r) = 0, \quad \forall r \le L,$
>
> *Growth:* $U(r) \ge a_0 r^\beta, \ |U'(r)| \le a_1 r^{\beta-1}, \ |U''(r)| \le a_2 r^{\beta-2}, \quad \forall r > L',$
>
> *Convexity:* $U'(r), U''(r) \ge 0, \quad \forall r > 0.$

In the range of parameters given by

$$\gamma < \begin{cases} 1, & 1 < \beta < \dfrac{4}{3}, \\[2mm] \dfrac{3}{2}\beta - 1, & \dfrac{4}{3} \le \beta < 2, \\[2mm] 2, & \beta \ge 2, \end{cases} \tag{3.16}$$

all solutions to system (3.14) *flock*

$$\mathcal{D}(t) \leq \overline{\mathcal{D}} < \infty$$

and align

$$\mathcal{E}(t) \leq \frac{C_\delta}{\langle t \rangle^{1-\delta}}, \qquad \forall \delta > 0. \tag{3.17}$$

Proof We will operate with the particle energy defined similarly to the confinement case:

$$\mathcal{E}_i = \frac{1}{2}|\mathbf{v}_i|^2 + \frac{1}{N}\sum_{k=1}^{N} U(\mathbf{x}_{ik}), \qquad \mathcal{E}_\infty = \max_i \mathcal{E}_i.$$

First, let us observe that the particle energy controls the diameter of the flock. By convexity and our assumptions on the growth of the potential, we have

$$\mathcal{E}_i \geq U(\mathbf{x}_i) \geq (|\mathbf{x}_i| - L')_+^\beta. \tag{3.18}$$

So

$$\mathcal{D} \leq \mathcal{E}_\infty^{1/\beta} + L'. \tag{3.19}$$

Let us now establish a bound on $\mathcal{E}_\infty$. For each i, we compute

$$\frac{d}{dt}\mathcal{E}_i = \frac{1}{N}\sum_{k=1}^{N}\phi_{ik}\mathbf{v}_{ki} \cdot \mathbf{v}_i - \frac{1}{N}\sum_{k=1}^{N}\nabla U(\mathbf{x}_{ik}) \cdot \mathbf{v}_k. \tag{3.20}$$

For the kinetic part, we use the identity

$$\mathbf{v}_{ki} \cdot \mathbf{v}_i = -\frac{1}{2}|\mathbf{v}_{ki}|^2 - \frac{1}{2}|\mathbf{v}_i|^2 + \frac{1}{2}|\mathbf{v}_k|^2. \tag{3.21}$$

Discarding all the negative terms, we bound

$$\frac{1}{N}\sum_{k=1}^{N}\phi_{ik}\mathbf{v}_{ki} \cdot \mathbf{v}_i \leq \|\phi\|_\infty \mathcal{K}.$$

Due to the energy law, $\mathcal{K}$, of course, will remain bounded, but we will keep it for now. As to the potential term, there are several ways we can handle it.

For any $1 \le \beta \le \frac{4}{3}$, we can derive a direct estimate from the first derivative:

$$\left| \frac{1}{N} \sum_{k=1}^{N} \nabla U(\mathbf{x}_{ik}) \cdot \mathbf{v}_k \right| \le \sqrt{\mathcal{K}} \left(\frac{1}{N} \sum_{k=1}^{N} |\nabla U(\mathbf{x}_{ik})|^2 \right)^{\frac{1}{2}} \le \sqrt{\mathcal{K}} \mathcal{D}^{\beta-1}.$$

Consequently,

$$\frac{\mathrm{d}}{\mathrm{d}t} \mathcal{E}^{(i)} \le c_1 \mathcal{K} + c_2 \sqrt{\mathcal{K}} \mathcal{D}^{\beta-1} \lesssim \sqrt{\mathcal{K}} (1 + \mathcal{E}_\infty^{\frac{\beta-1}{\beta}}),$$

and

$$\frac{\mathrm{d}}{\mathrm{d}t} \mathcal{E}_\infty \le c_3 \sqrt{\mathcal{K}} (1 + \mathcal{E}_\infty^{\frac{\beta-1}{\beta}}) \quad \Rightarrow \quad \mathcal{E}_\infty \lesssim \langle t \rangle^\beta \quad \Rightarrow \quad \mathcal{D} \lesssim \langle t \rangle. \tag{3.22}$$

In the range $\frac{4}{3} \le \beta \le 2$, it is better to make use of the second derivative:

$$\left| \frac{1}{N} \sum_{k=1}^{N} \nabla U(\mathbf{x}_{ik}) \cdot \mathbf{v}_k \right| = \frac{1}{N} \sum_{k=1}^{N} (\nabla U(\mathbf{x}_{ik}) - \nabla U(\mathbf{x}_i)) \cdot \mathbf{v}_k$$

$$\le \|D^2 U\|_\infty \sqrt{\mathcal{K}} \left(\frac{1}{N} \sum_{k=1}^{N} |\mathbf{x}_k|^2 \right)^{\frac{1}{2}} \tag{3.23}$$

$$\le c_4 \sqrt{\mathcal{K}} \left(\frac{1}{N^2} \sum_{i,j=1}^{N} |\mathbf{x}_{ij}|^2 \right)^{\frac{1}{2}}.$$

The following inequality will be used repeatedly:

$$\frac{1}{N^2} \sum_{i,j=1}^{N} |\mathbf{x}_{ij}|^2 \le (L')^2 + \frac{1}{N^2} \sum_{i,j=1}^{N} (|\mathbf{x}_{ij}| - L')_+^2 \le C(1 + \mathcal{D}^{(2-\beta)+} \mathcal{P}). \tag{3.24}$$

Continuing the above,

$$\left| \frac{1}{N} \sum_{k=1}^{N} \nabla U(\mathbf{x}_{ik}) \cdot \mathbf{v}_k \right| \le c_4 \sqrt{\mathcal{K}} (1 + \mathcal{D}^{2-\beta} \mathcal{P})^{1/2} \le c_5 \sqrt{\mathcal{K}} (1 + \mathcal{E}_\infty)^{\frac{2-\beta}{2\beta}}.$$

In this case,

$$\frac{d}{dt}\mathcal{E}_\infty \leq c_6\sqrt{\mathcal{K}}(1+\mathcal{E}_\infty)^{\frac{2-\beta}{2\beta}} \quad \Rightarrow \quad \mathcal{E}_\infty \lesssim \langle t\rangle^{\frac{2\beta}{3\beta-2}}$$

$$\Rightarrow \quad \mathcal{D} \leq \langle t\rangle^{\frac{2}{3\beta-2}}. \tag{3.25}$$

Finally, for $\beta > 2$, we argue similarly, using that $|D^2 U(\mathbf{x}_{ik})| \leq \mathcal{D}^{\beta-2}$, and (3.24), to obtain

$$\left|\frac{1}{N}\sum_{k=1}^N \nabla U(\mathbf{x}_{ik})\cdot\mathbf{v}_k\right| \leq \sqrt{\mathcal{K}}\mathcal{D}^{\beta-2},$$

and hence,

$$\frac{d}{dt}\mathcal{E}_\infty \leq c_7\sqrt{\mathcal{K}}(1+\mathcal{E}_\infty)^{\frac{\beta-2}{\beta}} \quad \Rightarrow \quad \mathcal{E}_\infty \lesssim \langle t\rangle^{\frac{\beta}{2}} \quad \Rightarrow \quad \mathcal{D} \leq \langle t\rangle^{\frac{1}{2}}. \tag{3.26}$$

We have proved the following a priori estimate:

$$\mathcal{D}(t) \lesssim \langle t\rangle^d, \quad \text{where} \quad d = \begin{cases} 1, & 1 \leq \beta < \frac{4}{3}, \\[2mm] \dfrac{2}{3\beta-2}, & \frac{4}{3} \leq \beta < 2, \\[2mm] \dfrac{1}{2}, & \beta \geq 2. \end{cases} \tag{3.27}$$

Denote

$$\zeta(t) = \langle t\rangle^{-\gamma d}.$$

According to the basic energy equation (3.15), we have

$$\frac{d}{dt}\mathcal{E} \leq -\frac{1}{2}\mathcal{I} - c\zeta(t)\mathcal{K}. \tag{3.28}$$

Considering this as a starting point, just like in the quadratic confinement case, we will build correctors to the energy to achieve full coercivity on the right-hand side of (3.28). We introduce one more auxiliary power function:

$$\eta(t) = \langle t\rangle^{-\alpha}, \quad \gamma d \leq \alpha < 1.$$

First, we consider the same longitudinal momentum:

$$\mathcal{X} = \frac{1}{N}\sum_{i=1}^N \mathbf{x}_i\cdot\mathbf{v}_i.$$

It will come with a prefactor $\varepsilon\eta(t)$, where ε is a small parameter. Let us estimate using (3.24):

$$\varepsilon\eta(t)|\mathcal{X}| \leq \varepsilon\mathcal{K} + \varepsilon\eta^2(t)\frac{1}{N^2}\sum_{i,j=1}^{N}|\mathbf{x}_{ij}|^2 \leq \varepsilon\mathcal{K} + c\varepsilon\eta^2(t) + \varepsilon\eta^2(t)\mathcal{D}^{(2-\beta)+}\mathcal{P}.$$

The potential term is bounded by $\varepsilon\mathcal{P}$ as long as

$$2\alpha \geq d(2-\beta)_+.$$

Hence,

$$\varepsilon\eta(t)|\mathcal{X}| \leq \varepsilon\mathcal{E} + c\eta^2(t). \tag{3.29}$$

This shows that

$$\mathcal{E} + \varepsilon\eta(t)\mathcal{X} + 2c\eta^2(t) \sim \mathcal{E} + c\varepsilon\eta^2(t).$$

Let us now consider the derivative

$$\mathcal{X}' = \frac{1}{N^2}\sum_{i=1}^{N}|\mathbf{v}_i|^2 + \frac{1}{N^2}\sum_{i,k=1}^{N}\mathbf{x}_{ik}\cdot\mathbf{v}_{ki}\phi_{ki} - \frac{1}{N^2}\sum_{i,k=1}^{N}\mathbf{x}_{ik}\cdot\nabla U(\mathbf{x}_{ik}) = \mathcal{K} + A - B.$$

The gain term B by convexity dominates the potential energy $B \geq \mathcal{P}$. This is the main reason why we introduced the $\mathcal{X}$-corrector. As to A,

$$|A| \leq \frac{\|\phi\|_\infty}{2\varepsilon^{1/2}\eta(t)}\mathcal{I} + \frac{\varepsilon^{1/2}\eta(t)}{2}\frac{1}{N^2}\sum_{i,j=1}^{N}|\mathbf{x}_{ij}|^2$$

$$\lesssim \frac{1}{\varepsilon^{1/2}\eta(t)}\mathcal{I} + \varepsilon^{1/2}\eta(t) + \varepsilon^{1/2}\eta(t)\mathcal{D}^{(2-\beta)+}\mathcal{P}.$$

By requiring a more stringent assumption on the parameters

$$\alpha \geq d(2-\beta)_+,$$

we can ensure that the potential term is bounded by $\sim \varepsilon^{1/2}\mathcal{P}$, which can be absorbed by the gain term. So far, we have obtained

$$\frac{d}{dt}(\mathcal{E} + \varepsilon\eta(t)\mathcal{X} + 2c\eta^2(t)) \leq -c_1\varepsilon\eta(t)\mathcal{E} + c_2\eta^2(t) + \varepsilon\eta'(t)\mathcal{X}. \tag{3.30}$$

In view of (3.29),

$$|\varepsilon \eta'(t)\mathcal{X}| \le \varepsilon \frac{1}{\langle t \rangle} \eta(t)|\mathcal{X}| \le \varepsilon \frac{1}{\langle t \rangle} \mathcal{E} + \varepsilon \frac{\eta^2(t)}{\langle t \rangle}.$$

Since $\alpha < 1$, the energy term will be absorbed, and the free term is even smaller than η^2. Denoting

$$E = \mathcal{E} + \varepsilon \eta(t)\mathcal{X} + 2c\eta^2(t),$$

we obtain

$$\frac{\mathrm{d}}{\mathrm{d}t}E \le -c_1 \eta(t)E + c_2 \eta^2(t).$$

By Duhamel's principle,

$$E(t) \lesssim \exp\{-\langle t \rangle^{1-\alpha}\} + \exp\{-\langle t \rangle^{1-\alpha}\} \int_0^t \frac{e^{\langle s \rangle^{1-\alpha}}}{\langle s \rangle^{2\alpha}} \, \mathrm{d}s.$$

By an elementary asymptotic analysis,

$$\int_0^t \frac{e^{\langle s \rangle^{1-\alpha'}}}{\langle s \rangle^{\alpha''}} \, \mathrm{d}s \sim \exp\{\langle t \rangle^{1-\alpha'}\} \frac{1}{\langle t \rangle^{\alpha''-\alpha'}}.$$

Thus, we obtain an algebraic decay rate:

$$E(t) \lesssim \frac{1}{\langle t \rangle^{\alpha}}, \quad \forall \alpha < 1, \tag{3.31}$$

provided

$$d\gamma < 1 \quad \text{and} \quad d(2 - \beta)_+ < 1. \tag{3.32}$$

This translates exactly into the conditions on γ given by (3.16), and (3.31) automatically implies (3.17)

Going back to estimates (3.22) and (3.25), but keeping the kinetic energy with its established decay, we obtain a new decay rate for the diameter:

$$\mathcal{D} \le C_\delta \langle t \rangle^{\frac{d}{2}+\delta}, \quad \forall \delta > 0.$$

At the next stage, we prove flocking: $\mathcal{D}(t) < \overline{\mathcal{D}}$. In order to achieve this, we return again to the particle energy estimates. Let us denote

$$\mathcal{P}_i = \frac{1}{N} \sum_{k=1}^{N} U(\mathbf{x}_{ik}), \quad \mathcal{I}_i = \frac{1}{N} \sum_{k=1}^{N} \phi_{ik}|\mathbf{v}_{ki}|^2, \quad \mathcal{X}_i = \mathbf{x}_i \cdot \mathbf{v}_i.$$

Using (3.20), (3.21), (3.23), and (3.24) and the fact that $\mathcal{D}^{(2-\beta)+}\mathcal{P}$ has a negative rate of decrease, we obtain

$$\frac{d}{dt}\mathcal{E}_i \leq \mathcal{K} - \frac{1}{2}\phi(\mathcal{D})|\mathbf{v}_i|^2 - \mathcal{I}_i + c\sqrt{\mathcal{K}} \lesssim -\frac{1}{2}\phi(\mathcal{D})|\mathbf{v}_i|^2 - \mathcal{I}_i + \frac{1}{\langle t\rangle^{\frac{1}{2}-\delta}}, \quad \forall \delta > 0.$$

In view of (3.32), we can pick α and δ such that

$$\frac{d\gamma}{2} + \delta\gamma < \frac{1}{2} - 2\delta < \alpha < \frac{1}{2} - \delta,$$

$$(2-\beta)_+ d + 2\delta(2-\beta)_+ < 2\alpha.$$

$$\tag{3.33}$$

We use as before the auxiliary rate function $\eta(t) = \langle t\rangle^{-\alpha}$. Let us estimate the corrector:

$$|\varepsilon\eta(t)\mathcal{X}_i| \leq \varepsilon|\mathbf{v}_i|^2 + \varepsilon\eta^2(t)|\mathbf{x}_i|^2 \leq \varepsilon|\mathbf{v}_i|^2 + \varepsilon\eta^2(t)\mathcal{D}^{2-\beta}\mathcal{P}_i + L^2\varepsilon\eta^2(t)$$

$$\leq \varepsilon|\mathbf{v}_i|^2 + c\varepsilon\mathcal{P}_i + L^2\varepsilon\eta^2(t).$$

So

$$E_i := \mathcal{E}_i + \varepsilon\eta(t)\mathcal{X}_i + 2L^2\varepsilon\eta^2(t) \sim \mathcal{E}_i + L^2\varepsilon\eta^2(t).$$

Differentiating,

$$\mathcal{X}_i' = |\mathbf{v}_i|^2 + \frac{1}{N}\sum_{k=1}^N \mathbf{x}_i \cdot \mathbf{v}_{ki}\phi_{ki} - \frac{1}{N}\sum_{k=1}^N \mathbf{x}_{ik} \cdot \nabla U(\mathbf{x}_{ik})$$

$$+ \frac{1}{N}\sum_{k=1}^N \mathbf{x}_k \cdot (\nabla U(\mathbf{x}_{ik}) - \nabla U(\mathbf{x}_i))$$

$$\leq |\mathbf{v}_i|^2 + \varepsilon^{1/2}\eta(t)|\mathbf{x}_i|^2 + \frac{1}{\varepsilon^{1/2}\eta(t)}\mathcal{I}_i - \mathcal{P}_i + \frac{1}{N^2}\sum_{l,k=1}^N |\mathbf{x}_{kl}|^2$$

$$\leq |\mathbf{v}_i|^2 + \varepsilon^{1/2}L^2\eta(t) + \varepsilon^{1/2}\mathcal{D}^{(2-\beta)+}\eta(t)\mathcal{P}_i + \frac{1}{\varepsilon^{1/2}\eta(t)}\mathcal{I}_i - \mathcal{P}_i + C.$$

In view of (3.33), we have $\varepsilon^{1/2}\mathcal{D}^{(2-\beta)+}\eta(t) \lesssim \varepsilon^{1/2}$. So the potential term is absorbed by $-\mathcal{P}_i$, and we continue to estimate

$$\leq |\mathbf{v}_i|^2 + \frac{1}{\eta(t)}\mathcal{I}_i - \frac{1}{2}\mathcal{P}_i + C.$$

Again, in view of (3.33), $\eta(t)$ decays faster than $\phi(\mathcal{D})$. Plugging it into the energy equation, we obtain

$$\frac{d}{dt}E_i \leq -\varepsilon\eta(t)E_i + \eta(t) + \sqrt{K} + \varepsilon\eta'(t)\mathcal{X}_i.$$

As before, $\varepsilon\eta'(t)\mathcal{X}$ is a lower-order term which is absorbed in the negative energy term and $+\eta^2$. So

$$\frac{d}{dt}E_i \leq -\varepsilon\eta(t)E_i + \eta(t) + \sqrt{K}.$$

By our choice of constants (3.33), $\sqrt{K}$ decays faster than $\eta(t)$. Hence,

$$\frac{d}{dt}E_i \lesssim -\varepsilon\eta(t)E_i + \eta(t).$$

This proves boundedness of E_i, and hence that of $\mathcal{E}_i + L^2\varepsilon\eta^2(t)$, and consequently, that of $\mathcal{E}_i$. In view of (3.19), this implies flocking:

$$\mathcal{D}(t) < \overline{\mathcal{D}}, \quad \forall t > 0.$$

$\square$

It is interesting to note that when the support of the potential spans the entire line, $L = 0$, and U lands at the origin with at least a quadratic touch:

$$U(r) \geq a_0 r^2, r < L', \tag{3.34}$$

then we can establish exponential alignment in terms of the energy $\mathcal{E}$. Indeed, since we already know that the diameter is bounded, the basic energy equation reads

$$\frac{d}{dt}\mathcal{E} \leq -c_0\mathcal{K} - \frac{1}{2}\mathcal{I}.$$

The momentum corrector needs only an ε-prefactor to satisfy the bound:

$$|\varepsilon\mathcal{X}| \leq \varepsilon\mathcal{K} + \varepsilon c\mathcal{P}.$$

This is due to the assumed quadratic order of the potential near the origin and, again, boundedness of the diameter. Hence, $\mathcal{E}+\varepsilon\mathcal{X} \sim \mathcal{E}$. The rest of the argument is similar to the confinement case. We obtain

$$\mathcal{X} \lesssim \mathcal{K} + \varepsilon^{1/2}\mathcal{P} + \frac{1}{\varepsilon^{1/2}}\mathcal{I} - \mathcal{P} \leq \mathcal{K} - \frac{1}{2}\mathcal{P}\frac{1}{\varepsilon^{1/2}}\mathcal{I}.$$

Thus,

$$\frac{d}{dt}(\mathcal{E} + \varepsilon\mathcal{X}) \le -c_1\mathcal{E} \sim -c_1(\mathcal{E} + \varepsilon\mathcal{X}).$$

This proves exponential decay of $\mathcal{E}$. Going further to consider the individual particle energies, we discover similar decays. Indeed, denoting by Exp any quantity that decays exponentially fast, we follow the same scheme to obtain

$$\frac{d}{dt}\mathcal{E}_i \le -c_1|\mathbf{v}_i|^2 - \frac{1}{2}\mathcal{I}_i + \text{Exp}.$$

In view of $|\mathbf{x}_i|^2 \lesssim \mathcal{P}_i$, we have

$$\varepsilon|\mathcal{X}_i| \le \varepsilon|\mathbf{v}_i|^2 + \varepsilon\mathcal{P}_i.$$

So $\mathcal{E}_i + \varepsilon\mathcal{X}_i \sim \mathcal{E}_i$. Proceeding further as in the proof, we obtain

$$\mathcal{X}_i' \lesssim |\mathbf{v}_i|^2 + \frac{1}{\varepsilon^{1/2}}\mathcal{I}_i - \frac{1}{2}\mathcal{P}_i.$$

Thus,

$$\frac{d}{dt}(\mathcal{E}_i + \varepsilon\mathcal{X}_i) \le -c_1(\mathcal{E}_i + \varepsilon\mathcal{X}_i) + \text{Exp}.$$

This establishes exponential decay for $\mathcal{E}_\infty$ and hence for the individual velocities. This also proves that $\mathcal{D}(t) = \text{Exp}$. So the alignment outcome here is exponential aggregation.

Theorem 3.5 *Let us assume that the support of the potential spans the entire space and* (3.34). *Then the solutions flock and align exponentially fast:*

$$\mathcal{D}(t) + \max_{i=1,\dots,N} |\mathbf{v}_i(t) - \bar{\mathbf{v}}| \le Ce^{-\delta t},$$

for some $C, \delta > 0$.

3.4 Dynamics Under Rayleigh Friction: Grassmannian Reduction

When the flock has tendency to adjust to a certain speed level, or energy level, inherent to all agents regardless of initial condition, the system has to include a Rayleigh-type friction force given by

$$\mathbf{F}_i = \sigma\mathbf{v}_i(1 - |\mathbf{v}_i|^p), \quad p > 0.$$

where $\sigma > 0$ is a strength parameter. So the system reads

$$\begin{cases} \dot{\mathbf{x}}_i = \mathbf{v}_i, \\ \dot{\mathbf{v}}_i = \lambda \displaystyle\sum_{j=1}^{N} m_j \phi(\mathbf{x}_i - \mathbf{x}_j)(\mathbf{v}_j - \mathbf{v}_i) + \sigma \mathbf{v}_i (1 - |\mathbf{v}_i|^p). \end{cases} \tag{3.35}$$

While the force pushes all magnitudes $|\mathbf{v}_i|$ toward the same value 1, it might in fact counteract the directional alignment forces to produce a nontrivial dynamics which depends on the relative strengths of the forces involved. The best way to illustrate this is by the following example.

Example 3.1 Let us assume that we have a global communication $\phi \equiv 1$ and consider a two-agent system on the line where $v = v_1 = -v_2 > 0$. Then we have the system

$$\dot{x} = v, \qquad \dot{v} = -\lambda v + \sigma v(1 - v^2).$$

The equation can be solved explicitly. If the Cucker-Smale communication is weak, $\lambda < \sigma$, then the solution is given by

$$v = \frac{\sqrt{1 - \frac{\lambda}{\sigma}}}{\sqrt{1 + c_0^2 e^{-2t(\sigma - \lambda)}}}.$$

So as we can see, even global communication is not sufficient to provide alignment in this case.

When $\lambda = \sigma$, we obtain

$$v(t) = \frac{v_0}{\sqrt{2\sigma t v_0^2 + 1}}.$$

Hence, the solution aligns to 0 and does so only algebraically fast. It clearly does not converge to the natural value $v = 1$. At the same time we can see that the agents diverge, $x(t) \sim \sqrt{t}$. So no flocking occurs either.

Finally, when $\lambda > \sigma$, we obtain a positive alignment result:

$$v = \frac{c_0 e^{(\sigma - \lambda)t} \sqrt{\frac{\lambda}{\sigma} - 1}}{\sqrt{1 - c_0^2 e^{2t(\sigma - \lambda)}}}.$$

So in this case, $v \to 0$ exponentially fast and flocking ensues.

This threshold condition on the absolute communication illustrates the general phenomenon.

Theorem 3.6 *Let $\phi_* = \inf \phi > 0$ and $\lambda \phi_* M > \sigma$. Then $\mathcal{A} \to 0$ exponentially fast, and either all $\mathbf{v}_i \to 0$ or there exists a vector $\bar{\mathbf{v}} \in \mathbb{S}^{n-1}$ to which all $\mathbf{v}_i$ converge exponentially fast.*

The theorem will be proved below after we establish a few general properties of system (3.35).

Let us note that the aligning to $\bar{\mathbf{v}} = 0$ may happen at an algebraic rate as we have seen in the example above. In the same example, we have seen a special placement of vectors occupying both sides of the real line. This is the basic reason why there is misalignment—friction pushes vectors to have absolute value 1, which, under this symmetric configuration, amounts to stretching vectors in the opposite directions countering the alignment force. If we restrict the initial configuration just avoiding this particular scenario, i.e., when all vectors $\mathbf{v}_i(0)$ are located on the side of a hyperplane, then a result similar to the classical heavy tail threshold can be established.

To address this systematically, let us first note a few general properties of the system. A couple of obvious "cons" is that the system does not conserve momentum or is Galilean invariant. On the positive side, the system has a partial *weak maximum principle*; that is, for any linear functional $\ell \in \mathbb{R}^n$, if all $\ell(\mathbf{v}_i(0)) \geq 0$ for all i, then $\ell(\mathbf{v}_i(t)) \geq 0$ for all time. Indeed, let us find an $i = i(t)$ such that $\ell(\mathbf{v}_i) = \min_j \ell(\mathbf{v}_j)$. Then by Rademacher's lemma we have

$$\frac{d}{dt}\ell(\mathbf{v}_i) = \sum_k m_k \phi_{ik}(\ell(\mathbf{v}_k) - \ell(\mathbf{v}_i)) + \sigma\ell(\mathbf{v}_i)(\theta_i - |\mathbf{v}_i|^p) \geq \sigma\ell(\mathbf{v}_i)(1 - |\mathbf{v}_i|^p).$$

Integrating, we obtain

$$\ell(\mathbf{v}_i) \geq \ell(\mathbf{v}_i(0))e^{c(t)} \geq 0.$$

An implication of this principle is invariance of any convex sector Σ defined by intersection of positive hyperplanes of any collection of functionals:

$$\Sigma_{\mathcal{F}} = \bigcap_{\ell \in \mathcal{F}} \{\mathbf{v} : \ell(\mathbf{v}) \geq 0\}.$$

An important family of solutions is solutions that lie on one side of a hyperplane $\ell(\mathbf{v}_i) > 0$. Let us note that if initially all velocities lie on one side of a hyperplane, then they actually belong to a slightly narrower sector defined by the span of the initial velocity vectors:

$$\mathbb{R}^+ \times \text{conv}\{\mathbf{v}_i\}_{i=1}^N \subset \{\mathbf{v} : \ell(\mathbf{v}) \geq \varepsilon|\mathbf{v}|\}.$$

By rotational invariance of the system, we can assume without loss of generality that the solution lies above the hyperplane $\Pi_n = \{x_n = 0\}$. Thus, we have

$$v_i^n \geq \varepsilon |\mathbf{v}_i|, \quad \forall i = 1, \ldots, N. \tag{3.36}$$

Definition 3.1 We call solutions satisfying (3.36) *sectorial*.

Let us now establish a few unconditional facts about the long-time behavior of the new system. First, let us compute the evolution of $|\mathbf{v}_+| = \max_i |\mathbf{v}_i|$. Via the usual maximizing functional approach, we obtain

$$\frac{\mathrm{d}}{\mathrm{d}t}|\mathbf{v}_+| \leq \sigma |\mathbf{v}_+|(1 - |\mathbf{v}_+|^P).$$

Solving this ordinary differential inequality (ODI) directly, we obtain

$$|\mathbf{v}_+|(t) \leq \frac{e^t}{\sqrt{c_0^2 + e^{2t}}},$$

where c_0 depends on the initial condition only. So it is clear that the friction acts strongly on large magnitudes to bring all vectors exponentially fast to the unit ball:

$$|\mathbf{v}_i(t)| \leq 1 + O(e^{-t}). \tag{3.37}$$

Lemma 3.1 *If* $\lambda \phi_* M > \sigma$, *then the system aligns and flocks exponentially fast:*

$$\mathcal{A} \leq \mathcal{A}_0 e^{(\sigma - \lambda \phi_* M)t}.$$

Proof We start with a traditional computation which leads to

$$\frac{\mathrm{d}}{\mathrm{d}t}\mathcal{A} \leq -\phi_* M \mathcal{A} + \sigma \ell[\mathbf{v}_i(1 - |\mathbf{v}_i|^P) - \mathbf{v}_j(1 - |\mathbf{v}_j|^P)], \tag{3.38}$$

where ℓ, i, and j are a maximizing triple for $\mathcal{A}$. Consider the functional

$$G(\mathbf{w}) = \mathbf{w}(1 - |\mathbf{w}|^P), \quad \text{with} \quad D_{\mathbf{w}}G(\mathbf{w}) = \mathrm{Id} - |\mathbf{w}|^P \mathrm{Id} - p|\mathbf{w}|^{P-2}\mathbf{w} \otimes \mathbf{w}.$$

Thus,

$$\mathbf{v}_i(1 - |\mathbf{v}_i|^P) - \mathbf{v}_j(1 - |\mathbf{v}_j|^P) = D_{\mathbf{w}}G(\mathbf{w})(\mathbf{v}_i - \mathbf{v}_j),$$

for some $\mathbf{w}$ on the segment $[\mathbf{v}_i, \mathbf{v}_j]$. Considering $\ell = \frac{\mathbf{v}_i - \mathbf{v}_j}{|\mathbf{v}_i - \mathbf{v}_j|}$, we can dismiss the entire negative definite part of $D_{\mathbf{w}}G$, with the remaining part being Id. Therefore,

$$\frac{\mathrm{d}}{\mathrm{d}t}\mathcal{A} \leq (\sigma - \phi_* M)\mathcal{A}.$$

This finishes the lemma. $\square$

Continuing under the same assumption on the communication strength, we can in fact deduce much more precise information about the long-time dynamics. Let us denote by $E = E(t)$ any exponentially decaying function. We have so far

$$\dot{\mathbf{v}}_i = \sigma \mathbf{v}_i (1 - |\mathbf{v}_i|^p) + E.$$

Multiplying by $p\mathbf{v}_i|\mathbf{v}_i|^{p-2}$ and denoting $y = |\mathbf{v}_i|^p$, we obtain the following ODE:

$$\dot{y} = p\sigma y(1 - y) + E. \tag{3.39}$$

Although the pure forceless logistic equation is easy to solve (all positive solutions converge to $\bar{\theta}$ or stay 0 if initially zero), the analysis of the forced ODE requires elaboration. Let us keep in mind that we have a solution y that is a priori nonnegative.

Lemma 3.2 *Any nonnegative solution to* (3.39) *either converges to 0 or to 1. In the latter case, convergence occurs exponentially fast.*

Proof Indeed, suppose y does not converge to 0. Then there exists a $\delta > 0$ for which there exists a sequence of times $t_1, t_2, \ldots \to \infty$ such that $y(t_i) > \delta$. For t large enough, we have

$$p\sigma\delta(1 - \delta) + E(t) > 0.$$

Therefore, starting from some time t^*, $y(t)$ will never cross δ again: $y(t) > \delta$, $t > t^*$. Solving

$$\frac{\mathrm{d}}{\mathrm{d}t}(1 - y) = -p\sigma y(1 - y) + E,$$

by Duhamel's principle, we obtain

$$1 - y = (1 - y(t^*)) \exp\left\{-p\sigma \int_{t^*}^t y(s)\,\mathrm{d}s\right\} + \int_{t^*}^t E(s) \exp\left\{-p\sigma \int_s^t y(\tau)\,\mathrm{d}\tau\right\}\,\mathrm{d}s.$$

So $|1 - y|$ is an exponentially decaying quantity. $\square$

Proof (Proof of Theorem 3.6) Since $\mathcal{A} \to 0$, we conclude from Lemma 3.2 that either all $\mathbf{v}_i \to 0$ or all $|\mathbf{v}_i| \to 1$ exponentially fast. In the latter case, we conclude that $\dot{\mathbf{v}}_i = E$, and hence, all $\mathbf{v}_i$'s must converge to a single vector on $\mathbb{S}^{n-1}$. We therefore have a full description of the dynamics under absolute communication, and Theorem 3.6 is proved. $\square$

It is easy to see that under the strong absolute communication assumption of Theorem 3.6, convergence to 0 can be eliminated as a collective outcome for sectorial solutions. However, the assumption of sectoriality allows us to extend the result to a much more general class of kernels with the classical heavy tail condition. So we now turn our attention to the study of sectorial solutions.

We can assume without loss of generality that our solution lies in a sector of opening $< \pi$ in the upper half-space, and as such, it satisfies (3.36). First, we establish a general bound from below.

Lemma 3.3 *All sectorial solutions stay bounded away from zero:*

$$\min_{i=1,\dots,N} |\mathbf{v}_i| \geq c_0,$$

for all time.

Proof Indeed, denote $v_i^n = \min_k v_k^n$. Then

$$\frac{d}{dt} v_i^n = \sum_k m_k \phi_{ik}(v_k^n - v_i^n) + \sigma v_i^n (1 - |\mathbf{v}_i|^p) \geq \sigma v_i^n (1 - \varepsilon^{-p}(v_i^n)^p).$$

Solving this logistic ODE, we can see that v_i^n remains bounded away from 0. This finishes the proof of the lemma. □

Denote $\tilde{\mathbf{r}} = \frac{\mathbf{r}}{|\mathbf{r}|}$. For the system at hand (3.35), or in fact for any system with a force parallel to velocity, $\mathbf{F}_i \times \mathbf{v}_i = 0$, we can write down the following equation for the direction vectors:

$$\frac{d}{dt} \tilde{\mathbf{v}}_i = \sum_{k=1}^{N} m_k \frac{|\mathbf{v}_k|}{|\mathbf{v}_i|} \phi_{ik}(\mathbb{I} - \tilde{\mathbf{v}}_i \otimes \tilde{\mathbf{v}}_i)\tilde{\mathbf{v}}_k, \tag{3.40}$$

and for the angles $\cos(\gamma_{ij}) = \tilde{\mathbf{v}}_i \cdot \tilde{\mathbf{v}}_j$:

$$\frac{d}{dt} \cos(\gamma_{ij}) = \sum_{k=1}^{N} m_k \frac{|\mathbf{v}_k|}{|\mathbf{v}_i|} \phi_{ik}(\cos(\gamma_{jk}) - \cos(\gamma_{ij})\cos(\gamma_{ik}))$$
$$+ \sum_{k=1}^{N} m_k \frac{|\mathbf{v}_k|}{|\mathbf{v}_j|} \phi_{jk}(\cos(\gamma_{ik}) - \cos(\gamma_{ij})\cos(\gamma_{jk})). \tag{3.41}$$

Let us note that this system in dimension 3 and higher does not have an explicit dissipative structure. However, in 2D, there is one: for a planar arrangement of three angles in the upper half plane in which γ_{ij} is the largest and $\gamma_{ij} < \pi - \delta$, we then have

$$\gamma_{ik} + \gamma_{jk} = \gamma_{ij} < \pi - \delta.$$

So then

$$\cos(\gamma_{jk}) - \cos(\gamma_{ij})\cos(\gamma_{ik}) = \cos(\gamma_{ij} - \gamma_{ik}) - \cos(\gamma_{ij})\cos(\gamma_{ik})$$
$$= \sin(\gamma_{ij})\sin(\gamma_{ik}) \geq 0.$$

Consequently, in view of the upper and lower bounds on the velocities,

$$\frac{d}{dt}\cos(\gamma_{ij}) \geq c\phi(\mathcal{D})\sum_{k=1}^{N} m_k(\cos(\gamma_{jk}) - \cos(\gamma_{ij})\cos(\gamma_{ik})$$
$$+ \cos(\gamma_{ik}) - \cos(\gamma_{ij})\cos(\gamma_{jk}))$$
$$= c\phi(\mathcal{D})\sum_{k=1}^{N} m_k(\cos(\gamma_{ik}) + \cos(\gamma_{jk}))(1 - \cos(\gamma_{ij})).$$

Now

$$\cos(\gamma_{ik}) + \cos(\gamma_{jk}) = 2\cos\left(\frac{\gamma_{ij}}{2}\right)\cos\left(\frac{\gamma_{ik} - \gamma_{jk}}{2}\right) \geq c_0$$

due to the sectorial limitation on the angles. So

$$\frac{d}{dt}\cos(\gamma_{ij}) \geq cM\phi(\mathcal{D})(1 - \cos(\gamma_{ij})). \tag{3.42}$$

To obtain this equation for arbitrary dimension, we apply a Grassmannian reduction method. To describe the method, let us fix an arbitrary 2D plane Π containing the x_n-axis, and let us consider a projection of (3.40) onto Π:

$$\dot{\mathbf{v}}_i^\Pi = \sum_{k=1}^{N} m_k\phi_{ik}(\mathbf{v}_k^\Pi - \mathbf{v}_i^\Pi) + \sigma(1 - |\mathbf{v}_i|^p)\mathbf{v}_i^\Pi. \tag{3.43}$$

Noting that the n-th coordinates of the projections remain the same as for the original vectors, all norms of $\mathbf{v}_i^\Pi$ remain bounded above and below. Let us now write the system for the unit vectors $\tilde{\mathbf{v}}_i^\Pi = \frac{\mathbf{v}_i^\Pi}{|\mathbf{v}_i^\Pi|}$:

$$\frac{d}{dt}\tilde{\mathbf{v}}_i^\Pi = \sum_{k=1}^{N} m_k\frac{|\mathbf{v}_k^\Pi|}{|\mathbf{v}_i^\Pi|}\phi_{ik}(\mathbb{I} - \tilde{\mathbf{v}}_i^\Pi \otimes \tilde{\mathbf{v}}_i^\Pi)\tilde{\mathbf{v}}_k^\Pi, \tag{3.44}$$

keeping in mind that ϕ_{ik} still depend on the original coordinates $\mathbf{x}_i - \mathbf{x}_k$. From (3.44), we deduce the same system for angles $\cos(\gamma_{ij}^\Pi) = \tilde{\mathbf{v}}_i^\Pi \cdot \tilde{\mathbf{v}}_j^\Pi$ as for the original variables:

$$\frac{d}{dt}\cos(\gamma_{ij}^{\Pi}) = \sum_{k=1}^{N} m_k \frac{|\mathbf{v}_k^{\Pi}|}{|\mathbf{v}_i^{\Pi}|} \phi_{ik}(\cos(\gamma_{jk}^{\Pi}) - \cos(\gamma_{ij}^{\Pi})\cos(\gamma_{ik}^{\Pi}))$$

$$+ \sum_{k=1}^{N} m_k \frac{|\mathbf{v}_k^{\Pi}|}{|\mathbf{v}_j^{\Pi}|} \phi_{jk}(\cos(\gamma_{ik}^{\Pi}) - \cos(\gamma_{ij}^{\Pi})\cos(\gamma_{jk}^{\Pi})). \tag{3.45}$$

Note that

$$\gamma^{2D} = \max_{\Pi \in \mathcal{G}(1,n-1),i,j} \gamma_{ij}^{\Pi} \geq \pi - \delta,$$

where $\mathcal{G}(1, n - 1)$ is the space of 2D planes containing x_n-axis, which can be identified as the compact 1-Grassmannian manifold of $\mathbb{R}^{n-1}$.

Taking now the minimum over Π, i, j, writing (3.45) for a minimizer triple, and invoking Rademacher's lemma we run the 2D computation above to system (3.45):

$$\frac{d}{dt}\cos(\gamma_{ij}^{\Pi}) \geq cM\phi(\mathcal{D})(1 - \cos(\gamma_{ij}^{\Pi})), \tag{3.46}$$

or

$$\frac{d}{dt}(1 - \cos(\gamma^{2D})) \leq -cM\phi(\mathcal{D})(1 - \cos(\gamma^{2D})). \tag{3.47}$$

Now, let us observe an elementary inequality:

$$\gamma = \max_{i,j} \gamma_{ij} \leq \gamma^{2D}. \tag{3.48}$$

Indeed, let $\gamma = \gamma_{ij}$. Consider the two-dimensional plane Π spanned by x_n-axis and $\tilde{\mathbf{v}}_i - \tilde{\mathbf{v}}_j$. Note that $\tilde{\mathbf{v}}_i - \tilde{\mathbf{v}}_j = (\tilde{\mathbf{v}}_i)^{\Pi} - (\tilde{\mathbf{v}}_j)^{\Pi}$, and the angle between $(\tilde{\mathbf{v}}_i)^{\Pi}$ and $(\tilde{\mathbf{v}}_j)^{\Pi}$ is the same γ_{ij}^{Π}. So considering the two isosceles triangles spanned on $\tilde{\mathbf{v}}_i, \tilde{\mathbf{v}}_j$ and $(\tilde{\mathbf{v}}_i)^{\Pi}, (\tilde{\mathbf{v}}_j)^{\Pi}$ and applying the cosine theorem, we have

$$2(1 - \cos(\gamma)) = |\tilde{\mathbf{v}}_i - \tilde{\mathbf{v}}_j|^2 = 2|(\tilde{\mathbf{v}}_i)^{\Pi}|^2(1 - \cos(\gamma_{ij}^{\Pi}))$$

$$\leq 2(1 - \cos(\gamma_{ij}^{\Pi})) \leq 2(1 - \cos(\gamma^{2D})).$$

This proves (3.48). Let us note in passing, although we will not need it for future, that the opposite inequality is also true:

$$c\gamma^{2D} \leq \gamma,$$

where $0 < c \leq 1$ depends on the opening of the sector.

Coming back now to the original system, we derive one more equation for

$$\mathcal{R} = \max_{i',i''} \frac{|\mathbf{v}_{i''}|^2}{|\mathbf{v}_{i'}|^2} = \frac{|\mathbf{v}_{i(t)}|^2}{|\mathbf{v}_{j(t)}|^2}.$$

Note that $\mathcal{R}$ is a priori bounded from above and below. Using the usual Rademacher's lemma we write

$$\frac{d}{dt}\mathcal{R} = \frac{1}{|\mathbf{v}_j|^2} \sum_{k=1}^{N} m_k \phi_{ik}(\mathbf{v}_k \cdot \mathbf{v}_i - |\mathbf{v}_i|^2) + \frac{|\mathbf{v}_i|^2}{|\mathbf{v}_j|^4} \sum_{k=1}^{N} m_k \phi_{jk}(|\mathbf{v}_j|^2 - \mathbf{v}_k \cdot \mathbf{v}_j)$$
$$+ \mathcal{R}(|\mathbf{v}_j|^p - |\mathbf{v}_i|^p).$$

Note that the first sum is negative, so we simply dismiss it. In the second sum, we use

$$|\mathbf{v}_j|^2 - \mathbf{v}_k \cdot \mathbf{v}_j \le |\mathbf{v}_j|^2 - |\mathbf{v}_k||\mathbf{v}_j|\cos(\gamma) \le |\mathbf{v}_j|^2(1 - \cos(\gamma)) \lesssim (1 - \cos(\gamma)).$$

For the friction term, we observe

$$\mathcal{R}(|\mathbf{v}_j|^p - |\mathbf{v}_i|^p) \lesssim (1 - \mathcal{R}^{p/2}) \lesssim (1 - \mathcal{R}).$$

Thus,

$$\frac{d}{dt}(\mathcal{R} - 1) \le c_1(1 - \cos(\gamma)) - c_2(\mathcal{R} - 1). \tag{3.49}$$

This leads to our main result.

Theorem 3.7 *Suppose the kernel satisfies*

$$\phi(r) \ge \frac{1}{\langle t \rangle^\beta}, \quad \beta \le 1. \tag{3.50}$$

Then every sectorial solution to (3.35) aligns and flocks exponentially fast to $\bar{\mathbf{v}} \in \mathbb{S}^{n-1}$.

Proof We will use a bootstrap argument. Suppose $\beta < 1$ first. Since all velocities remain bounded, we have

$$\mathcal{D}(t) \lesssim t.$$

Using this in (3.47) and (3.48), we obtain

$$(1 - \cos(\gamma)) \lesssim e^{-c\langle t \rangle^{1-\beta}}.$$

Plugging this into the $\mathcal{R}$-equation (3.49) and using Duhamel's principle, we have

$$(\mathcal{R} - 1) \lesssim e^{-c\langle t\rangle^{1-\beta}}.$$

Noting the bound

$$\frac{d}{dt}\mathcal{D} \leq \mathcal{A} \lesssim \sqrt{(\mathcal{R} - 1) + (1 - \cos(\gamma))} \lesssim e^{-c\langle t\rangle^{1-\beta}},$$

we now see that the diameter of the flock remains bounded $\mathcal{D} \leq \overline{\mathcal{D}}$. Going back to systems (3.47)–(3.49), we conclude exponential decay for $(\mathcal{R} - 1) + (1 - \cos(\gamma))$ and hence for $\mathcal{A}$.

Next, with the obtained information, we can write the equations for extreme norms:

$$\frac{d}{dt}|\mathbf{v}_\pm|^p = p\sigma|\mathbf{v}_\pm|^p(1 - |\mathbf{v}_\pm|^p) + E(t).$$

Since $c \leq |\mathbf{v}_\pm|^p \leq C$, we conclude that $1 - |\mathbf{v}_\pm|^p$ tend to zero exponentially fast. This immediately implies that

$$\frac{d}{dt}\mathbf{v}_i = E, \quad \forall i,$$

and hence, each vector has a limit, which is common for all $\mathbf{v}_i$, as $t \to \infty$. This finishes the proof for the case of $\beta < 1$.

Let us turn to the case $\beta = 1$. Here, we make one more preliminary step: from $\mathcal{D} \lesssim t$, we deduce that

$$(1 - \cos(\gamma)) \lesssim \frac{1}{\langle t\rangle^\eta}, \quad \eta > 0.$$

Hence, due to asymptotic formula $e^{-ct} * \frac{1}{\langle t\rangle^\eta} \sim \frac{1}{\langle t\rangle^\eta}$,

$$\mathcal{R} - 1 \lesssim \frac{1}{\langle t\rangle^\eta}.$$

This in turn implies

$$\mathcal{D} \lesssim \langle t\rangle^{1-\frac{\eta}{2}}.$$

Plugging into the kernel again, this is essentially the same as to assume that $\beta < 1$, and the previous steps get repeated to finish the theorem. □

Example 3.2 A modification of Example 2.1 shows that the heavy tail condition $\beta \leq 1$ is necessary. Indeed, let us consider a two-agent system with $\mathbf{v}' = \langle v_1, v_2\rangle$ and $\mathbf{v}'' = \langle -v_1, v_2\rangle$. We assume that the kernel is given by the exact power law

$\phi(r) = \frac{1}{r^\beta}$ for $r > r_0$. The initial condition for the coordinates of the agents $\mathbf{x}' = \langle x_1, x_2 \rangle$, $\mathbf{x}'' = \langle -x_1, x_2 \rangle$ is such that $x_1(0) > 2r_0$. Then as long as $x_1(t) > r_0$, we have

$$\begin{cases} \dfrac{d}{dt} v_1 = -\dfrac{v_1}{x_1^\beta} + \sigma v_1 (1 - |\mathbf{v}'|^p), \\[4mm] \dfrac{d}{dt} v_2 = \sigma v_2 (1 - |\mathbf{v}'|^p). \end{cases}$$

Now, if $|\mathbf{v}'(0)| < 1$, it will remain so by the maximum principle. Then we obtain the system

$$\frac{d}{dt} v_1 \geq -\frac{v_1}{x_1^\beta}, \quad \frac{d}{dt} x_1 = v_1.$$

The system has a Lyapunov function:

$$L = v_1 + \frac{x_1^{1-\beta}}{1-\beta},$$

which decays along the trajectories. Thus, since $\beta > 1$,

$$v_1(t) \geq L(t) \geq v_1(0) + \frac{x_1^{1-\beta}(0)}{1-\beta}.$$

So if r_0 is sufficiently small, we can set $1 > v_1(0) > \frac{x_1^{1-\beta}(0)}{\beta-1}$, and $v_2(0)$ is small too to satisfy $|\mathbf{v}'(0)| < 1$. The above computation then shows that $v_1(t) > c_0 > 0$. This in part implies that $x_1(t)$ is increasing, and hence, the condition $x_1 > r_0$ will hold indefinitely. Hence, $v_1(t) > c_0$ holds indefinitely too. This establishes misalignment.

3.5   Notes and References

The first version of Theorem 3.1 was proved by Kim and Peszek in [61] for the case of a quadratic attraction/repulsion potential $U(r) = (r - R)^2$ and non-degenerate communication ϕ. Even though the potential is not singular at the origin, one can still prove no collisions starting from some time T. Kinetic repulsion force was introduced by Cucker and Dong in [31] and in [32] generalized to nonlinear couplings.

The N-independent analysis of 2-zone attraction models and the hypocoercivity method was introduced by Shu and Tadmor in recent works [90, 91]. The latter in fact addresses a more general class of Hamiltonian systems with anticipation which leads to systems of Cucker-Smale type with matrix communication. We essentially follow their work with minor modifications which allow to include a wider range of indexes β, γ, potentials extended to act from scale L, and Theorem 3.5 going one step further by establishing exponential flocking in the case $L = 0$.

The problem which still remains unexplored in the context of repulsion/alignment or 3-zone models is whether one can establish some flocking behavior for large crowds $N \to \infty$. This either amounts to finding N-independent estimates or working directly with the kinetic or hydrodynamic systems, which are covered in the corresponding chapters below.

The system with Rayleigh friction forcing, and $p = 2$, was first studied by Ha et al. in [46], where a prototype of Theorem 3.6 in the ℓ^2-framework was established along with alignment for solutions confined to a coordinate sector. The extension presented in Theorem 3.7 and the method of Grassmannian reduction appeared in [67].

Chapter 4
Kinetic Models

In the large crowd systems, where $N \sim \infty$, it is more efficient to resort to the mesoscopic level of description of the Cucker-Smale dynamics. The corresponding kinetic formulation of (2.1) can be derived formally via the BBGKY hierarchy. Looking slightly ahead, we seek to derive the following Vlasov-type model which describes evolution of a mass probability distribution $f(x, v, t)$ of agents in phase space (x, v):

$$\partial_t f + v \cdot \nabla_x f + \lambda \nabla_v \cdot [f F(f)] = 0, \qquad (4.1)$$

where

$$F(f)(x, v, t) = \int_{\mathbb{R}^{2n}} \phi(x, y)(w - v) f(y, w, t) \, dw \, dy.$$

We will present a rigorous derivation of (4.1) via the mean-field limit and along the way establish contractivity and asymptotic stability estimates for solutions.

4.1 BBGKY Hierarchy: Formal Derivation

Let us consider a probability density

$$P^N = P^N(x_1, v_1, \ldots, x_N, v_N, t)$$

of a system of N agents in the ensemble configuration space:

$$(x_1, v_1, \ldots, x_N, v_N) \in \mathbb{R}^{2nN}.$$

© Springer Nature Switzerland AG 2021
R. Shvydkoy, *Dynamics and Analysis of Alignment Models of Collective Behavior*,
Nečas Center Series, https://doi.org/10.1007/978-3-030-68147-0_4

The conservation of mass in the Gibbs ensemble propagated according to the given system (2.1) leads to the classical Liouville equation:

$$P_t^N + \sum_{i=1}^N v_i \cdot \nabla_{x_i} P^N + \sum_{i=1}^N \nabla_{v_i} \cdot (\dot{v}_i P^N) = 0. \tag{4.2}$$

We assume the effective radius of communication between agents remains independent of N, i.e., the kernel ϕ is not rescaled with N. This scaling regime is called *the mean-field limit*. We further assume that the total mass $M = \sum m_i$ remains constant and $\max_i m_i \to 0$. As a result, the agents become more and more indistinguishable, which we reflect in the symmetry condition:

$$P^N(\dots, x_i, v_i, \dots, x_j, v_j, \dots, t) = P^N(\dots, x_j, v_j, \dots, x_i, v_i, \dots, t).$$

We seek to derive an equation for the first marginal:

$$P^{1,N}(x, v, t) = \int_{\mathbb{R}^{2n(N-1)}} P^N(x, v, \bar{x}, \bar{v}, t) d\bar{x} d\bar{v},$$

where $\bar{x} = (x_2, \dots, x_N)$ and $\bar{v} = (v_2, \dots, v_N)$. Thus, integrating in $\bar{x}, \bar{v}$ in (4.2), we obtain

$$P_t^{1,N} + v \cdot \nabla_x P^{1,N} + \lambda \nabla_v \cdot \int_{\mathbb{R}^{2n(N-1)}} \sum_{j=2}^N m_j \phi(x, x_j)(v_j - v) P^N d\bar{x} d\bar{v} = 0.$$

In view of the symmetry of P^N, we achieve equality of the integrals in the sum above, and hence,

$$P_t^{1,N} + v \cdot \nabla_x P^{1,N}$$
$$+ \lambda(M - m_1)\nabla_v \cdot \int_{\mathbb{R}^{2n}} \phi(x, y)(w - v) P^{2,N}(x, v, y, w, t) \, dy \, dw = 0,$$

where $P^{2,N}$ is the second marginal:

$$P^{2,N}(x, v, y, w, t) = \int_{\mathbb{R}^{2n(N-2)}} P^N(x, v, y, w, \bar{\bar{z}}, \bar{\bar{u}}, t) d\bar{\bar{z}} d\bar{\bar{u}}.$$

Denoting the limiting densities by $P = \lim_{N \to \infty} P^{1,N}$, $Q = \lim_{N \to \infty} P^{2,N}$, we obtain

$$P_t + v \cdot \nabla_x P + \lambda M \nabla_v \cdot \int_{\mathbb{R}^{2n}} \phi(x, y)(w - v) Q(x, v, y, w, t) \, dy \, dw = 0.$$

We close by making the molecular chaos assumption:

$$Q(x, v, y, w, t) = P(x, v, t)P(y, w, t),$$

which results in precisely the following Vlasov-type equation (4.1) for the mass density $f = MP$.

4.2 Weak Formulation and Basic Principles of Kinetic Dynamics

We now turn to rigorous study and justification of the kinetic model (4.1). For simplicity, this will be done under the assumption that the kernel ϕ is of convolution type, decreasing and smooth, $\phi \in C^1$.

A straightforward connection between discrete and kinetic models can be seen by considering the empirical measure:

$$\mu_t^N = \sum_{i=1}^{N} m_i \delta_{\mathbf{x}_i(t)} \otimes \delta_{\mathbf{v}_i(t)}, \tag{4.3}$$

which satisfies a weak formulation of (4.1) if and only if the system $\{\mathbf{x}_i, \mathbf{v}_i\}_i$ solves the classical Cucker-Smale equations (2.1). This leads us first to define (4.1) in a weak sense for a special class of measure-valued solutions. We denote by $\mathcal{M}_+(\mathbb{R}^{2n})$ the set of nonnegative Radon measures on $\mathbb{R}^{2n}$, and

$$\mathcal{M}_+^p = \{\mu \in \mathcal{M}_+(\mathbb{R}^{2n}) : \|\mu\|_p = \int_{\mathbb{R}^{2n}} (1 + |v|^p)\, d\mu(x, v) < \infty\}.$$

We endow $\mathcal{M}_+$ and $\mathcal{M}_+^p$ with the topology of weak convergence, which means convergence on continuous bounded function $C_b(\mathbb{R}^{2n})$. Note that any bounded set in $\mathcal{M}_+^2$ is automatically tight; that is, for every $\varepsilon > 0$, there exists a compact set K_ε (a ball in fact) such that $\mu(\mathbb{R}^{2n} \setminus K_\varepsilon) < \varepsilon$ for all μ in the family. So by Prohorov's theorem, weak convergence on $C_b(\mathbb{R}^{2n})$ is equivalent to the classical weak* convergence on $C_0(\mathbb{R}^{2n})$, the predual of $\mathcal{M}(\mathbb{R}^{2n})$. Therefore, all bounded sets in $\mathcal{M}_+^2$ are weakly precompact. Since we will deal with measures confined to a bounded set anyway, we will not detail the statements above here.

Let us note that for any $\mu \in \mathcal{M}_+^p$, $p > 1$, the integral

$$F(\mu)(x, v) = \int_{\mathbb{R}^{2n}} \phi(x - y)(w - v)\, d\mu(y, w)$$

defines a C^1 smooth locally bounded field on $\mathbb{R}^{2n}$. Moreover, for a time-dependent weakly continuous family $\{\mu_t\}_{0 \le t < T} \in C_{w*}([0, T); \mathcal{M}_+^p(\mathbb{R}^{2n}))$, $p > 1$, the field

F becomes continuous $F(\mu.) \in C([0, T) \times \mathbb{R}^{2n})$ and uniformly Lipschitz $F(\mu.) \in L^{\infty}([0, T); \text{Lip}_{\text{loc}}(\mathbb{R}^{2n}))$ with

$$|F(\mu_t)(x, v)| \leq C(1 + |v|). \tag{4.4}$$

This is sufficient to define the proper global flow map on $[0, T) \times \mathbb{R}^{2n}$ later.

Definition 4.1 We say that $\{\mu_t\}_{0 \leq t < T} \in C_{w*}([0, T); \mathcal{M}_+^2(\mathbb{R}^{2n}))$ is a measure-valued solution to (4.1) with initial condition μ_0 if for any test-function $g \in C_0^{\infty}([0, T) \times \mathbb{R}^{2n})$ one has, for all $0 < t < T$,

$$\int_{\mathbb{R}^{2n}} g(t, x, v) \, d\mu_t(x, v) = \int_{\mathbb{R}^{2n}} g(0, x, v) \, d\mu_0(x, v)$$

$$+ \int_0^t \int_{\mathbb{R}^{2n}} (\partial_s g + v \cdot \nabla_x g + \lambda F(\mu_s) \cdot \nabla_v g) \, d\mu_s(x, v) \, ds. \tag{4.5}$$

The proof of the following lemma is straightforward.

Lemma 4.1 *The empirical measure* (4.3) *satisfies the weak kinetic formulation* (4.5) *if and only if the set of pairs* $(\mathbf{x}_i, \mathbf{v}_i)_i$ *is a solution to* (2.1).

The choice of $p = 2$ is motivated by our interest in the energy class; however, any $p > 1$ is sufficient to make the right-hand side of (4.5) well-defined. From compactly supported test functions, formulation (4.5) is easily extendable to functions with quadratic growth of the material derivative: $|\partial_s g| + |v||\nabla_x g| + |v||\nabla_v g| \lesssim |v|^2$. Immediate consequences are the mass conservation (plugging $g = 1$):

$$\frac{d}{dt} \mu_t(\mathbb{R}^{2n}) = 0, \tag{4.6}$$

conservation of momentum (plugging $g = v_i$),

$$\frac{d}{dt} \int_{\mathbb{R}^{2n}} v \, d\mu_t(x, v) = 0, \tag{4.7}$$

and the energy law (plugging $g = |v|^2$),

$$\frac{d}{dt} \int_{\mathbb{R}^{2n}} \frac{1}{2} |v|^2 \, d\mu_t(x, v) = -\frac{\lambda}{2} \int_{\mathbb{R}^{2n}} \int_{\mathbb{R}^{2n}} \phi(x - y)|w - v|^2 \, d\mu_t(y, w) \, d\mu_t(x, v).$$

One implication of the decaying energy is that any solution will automatically retain uniformly bounded second momentum on its interval of existence:

$$\|\mu_t\|_2 \leq \|\mu_0\|_2. \tag{4.8}$$

Another easy application is concentration toward macroscopic values in L^2-sense, which presents a direct analogue of the discrete result (2.1). To make this precise, let us denote the average velocity:

$$\overline{V} = \frac{1}{M} \int_{\mathbb{R}^{2n}} v \, d\mu_t(x, v).$$

It determines the direction of the flock centered around the mass center:

$$\overline{X} = \int_{\mathbb{R}^{2n}} x \, d\mu_t(x, v), \qquad \frac{d}{dt}\overline{X} = \overline{V}.$$

Due to Galilean invariance of the Eq. (4.1),

$$\mu_t \rightarrow \widetilde{\mu}_t, \qquad \int g(t, x, v)\widetilde{\mu}_t(x, v) = \int g(t, x + \overline{X}(t), v + \overline{V}) \, d\mu_t(x, v)$$

we can always assume that $\overline{V} = \overline{X} = 0$. Assuming for the sake of simpler discussion that μ_t are given by density distributions $f(x, v, t)$, we consider the macroscopic quantities:

$$\rho(x, t) = \int_{\mathbb{R}^n} f(x, v, t) \, dv, \qquad \rho\mathbf{u} = \int_{\mathbb{R}^n} v f(x, v, t) \, dv. \tag{4.9}$$

Then the total kinetic energy can be decomposed into internal (peculiar) and macroscopic energies as follows:

$$\mathcal{E} = \frac{1}{2} \int_{\mathbb{R}^{2n}} |v|^2 f(x, v, t) \, dx \, dv = \mathcal{E}_I + \mathcal{E}_m,$$

$$\mathcal{E}_I = \frac{1}{2} \int_{\mathbb{R}^{2n}} |v - \mathbf{u}(x, t)|^2 f(x, v, t) \, dx \, dv, \qquad \mathcal{E}_m = \frac{1}{2} \int_{\mathbb{R}^n} \rho(x, t)|\mathbf{u}(x, t)|^2 \, dx.$$

Note that due to the assumed zero momentum condition we have

$$\mathcal{E}_m = \frac{1}{4M} \int_{\mathbb{R}^{2n}} \rho(x, t)\rho(y, t)|\mathbf{u}(x, t) - \mathbf{u}(y, t)|^2 \, dx \, dy.$$

From (4.6), we obtain

$$\frac{d}{dt}\mathcal{E} \leq -2\phi(\mathcal{D}(t))M\mathcal{E},$$

where $\mathcal{D} = \sup_{x, y \in \text{Supp} f} |x - y|$. Arguing as in Sect. 2.4, we obtain a direct analogue of the discrete statement in L^2-terms:

$$\sup_{t \geq 0} \mathcal{D}(t) \leq \overline{\mathcal{D}}, \qquad \mathcal{E}(t) \leq \mathcal{E}_0 e^{-2t\lambda M\phi(\overline{\mathcal{D}})}. \tag{4.10}$$

Given the decomposition of the energy, this expresses both tendency of the distribution f to the macroscopic values and global alignment of the macroscopic velocity $\mathbf{u}$ on the support of the flock. Remarkably, the latter is derived without any knowledge of what macroscopic system $\mathbf{u}$ satisfies! We will discuss the closure problem in more detail later.

To fully exploit the transport structure of the kinetic equation (4.1), let us consider the characteristic flow of the field $\langle v, \lambda F(\mu_t) \rangle$:

$$\frac{\mathrm{d}}{\mathrm{d}t} X(t, s, x, v) = V(t, s, x, v), \qquad X(s, s, x, v) = x, \tag{4.11}$$

$$\frac{\mathrm{d}}{\mathrm{d}t} V(t, s, x, v) = \lambda F(\mu_t)(X, V), \quad V(s, s, x, v) = v. \tag{4.12}$$

We also denote $X(t, 0, x, v) = X(t, x, v)$, $V(t, 0, x, v) = V(t, x, v)$, and sometimes $(x, v) = \omega$. Note that $F(\mu_t)$ is smooth in ω, so the flow is well-defined and smooth. In view of the linear bound (4.4), we also conclude that the ODE (4.11)–(4.12) is well-posed on the entire existence time interval of the solution μ. Using the test function $g(s, \omega) = h(X(t, s, \omega), V(t, s, \omega))$ in (4.5), for some $h \in C_0^\infty(\mathbb{R}^{2n})$, we have

$$\partial_s g + v \cdot \nabla_x g + \lambda F(\mu_s) \cdot \nabla_v g = 0.$$

So (4.5) reads

$$\int_{\mathbb{R}^{2n}} h(\omega) \, \mathrm{d}\mu_t(\omega) = \int_{\mathbb{R}^{2n}} h(X(t, \omega), V(t, \omega)) \, \mathrm{d}\mu_0(\omega). \tag{4.13}$$

We say that μ_t is a *push-forward* of the measure μ_0 under the flow map (X, V), $\mu_t = (X, V)\#\mu_0$. This will be the key formula to deduce contractivity of the solution map to (4.1) and to study the mean-field limit. Due to finite second momenta of the measures, (4.13) is extendable to $|h| \lesssim |v|^2$ (note here a rough estimate on the velocity flow $|V| \leq C(t, s)(1 + |v|)$). Hence, we can apply (4.13) to the V-equation (4.12) to rewrite it completely in Lagrangian coordinates:

$$\frac{\mathrm{d}}{\mathrm{d}t} V(t, \omega) = \int_{\mathbb{R}^{2n}} \phi(X(t, \omega) - X(t, \omega'))(V(t, \omega') - V(t, \omega)) \, \mathrm{d}\mu_0(\omega'). \tag{4.14}$$

In this form, it presents a direct analogue to the discrete system (2.1), and as a result, we can carry out all the basic flocking results in a way very similar to the discrete case.

4.3 Kinetic Maximum Principle and Flocking

It is clear from (4.14) that the velocity characteristics are concentrating toward the mean value conserved in time. In a desire to quantify the rate of convergence, one stumbles upon the need for global communication in a way similar to the discrete case. Let us assume for now that we have a general nonnegative kernel $\phi = \phi(x - y)$ and work out a system of equations for the kinetic flock parameters. We define the amplitude and diameter over a general compact domain Ω containing the initial flock, $\text{Supp}\,\mu_0 \subset \Omega$. This will serve multiple purposes—to show alignment on an arbitrarily wide domain and to provide a tool to compare two different solutions. So let us assume that we are working with a given solution $\mu \in C_{w^*}([0, T); \mathcal{M}_+^2(\mathbb{R}^{2n}))$ with finite initial support. We define

$$\mathcal{D}_\Omega(t) = \max_{\omega', \omega'' \in \Omega} |X(t, \omega') - X(t, \omega'')|, \tag{4.15}$$

$$\mathcal{A}_\Omega(t) = \max_{\omega', \omega'' \in \Omega} |V(t, \omega') - V(t, \omega'')|. \tag{4.16}$$

We will perform a computation similar in spirit to the discrete system (2.23), but minding the wider range of the flock parameters. So let us pick a triple $\ell \in (\mathbb{R}^{2n})^*$, $|\ell| = 1$, $\omega', \omega'' \in \Omega$ which maximizes $\mathcal{A}_\Omega(t)$:

$$\mathcal{A}_\Omega(t) = \ell(V(t, \omega'') - V(t, \omega')).$$

We abbreviate $V(t, \omega) = V$, $V(t, \omega') = V'$, $V(t, \omega'') = V''$, etc. Then by the same argument as in discrete case,

$$\frac{\mathrm{d}}{\mathrm{d}t}\mathcal{A}_\Omega(t) = \lambda \int_{\mathbb{R}^{2n}} \phi(X - X'')\ell(V - V'') \,\mathrm{d}\mu_0(\omega)$$

$$+ \lambda \int_{\mathbb{R}^{2n}} \phi(X - X')\ell(V' - V) \,\mathrm{d}\mu_0(\omega)$$

$$\leq \lambda \phi(\mathcal{D}_\Omega) \int_{\mathbb{R}^{2n}} \ell(V - V'' + V' - V) \,\mathrm{d}\mu_0(\omega)$$

$$= -\lambda M \phi(\mathcal{D}_\Omega)\mathcal{A}_\Omega(t).$$

We also have trivially

$$\frac{\mathrm{d}}{\mathrm{d}t}\mathcal{D}_\Omega \leq \mathcal{A}_\Omega.$$

So we recover the same system of ordinary differential inequalities (ODIs) as in the discrete case (2.24). For general kernels, it simply implies that $\mathcal{A}_\Omega$ is a decreasing quantity:

$$\mathcal{A}_\Omega(t) \le \mathcal{A}_\Omega(0), \tag{4.17}$$

and hence, $\mathcal{D}_\Omega(t) \lesssim t$. In particular, this implies an a priori linear bound on the radius of the kinetic flock:

$$\operatorname{Supp}\mu_t \subset B_{R_1+tR_2}(0). \tag{4.18}$$

In the case of a heavy tail kernel, we obtain the full analogue of Theorem 2.2 in terms of kinetic parameters.

Theorem 4.1 (Alignment for Kinetic Model) *Let* $\mu \in C_{w^*}(\mathbb{R}^+; \mathcal{M}_+^2(\mathbb{R}^{2n}))$ *be a given solution with compact initial support and let* Ω *be a compact domain with* $\operatorname{Supp}\mu_0 \subset \Omega$. *Suppose* $\overline{\mathcal{D}}_\Omega$ *is a solution to*

$$\int_{\mathcal{D}_\Omega(0)}^{\overline{\mathcal{D}}_\Omega} \phi(r)\,dr = \frac{\mathcal{A}_\Omega(0)}{\lambda M}. \tag{4.19}$$

Then the solution μ *flocks exponentially fast according to*

$$\sup_{t\ge0} \mathcal{D}_\Omega(t) \le \overline{\mathcal{D}}_\Omega, \quad \mathcal{A}_\Omega(t) \le \mathcal{A}_\Omega(0)e^{-t\lambda M\phi(\overline{\mathcal{D}}_\Omega)}. \tag{4.20}$$

Besides flocking behavior, the estimates (4.20) imply that in the fat kernel case the supports of μ_t remain a priori uniformly bounded:

$$\operatorname{Supp}\mu_t \subset B_R(0), \quad \forall t \ge 0.$$

Further refinement of flocking behavior can be provided by establishing estimates on the deformation tensor of the flow map (X, V). Straight from the Lagrangian formulation, we obtain the system for all $t \ge 0$, $\omega \in \Omega$:

$$\partial_t \nabla X(t, \omega) = \nabla V(t, \omega)$$

$$\begin{aligned}
\partial_t \nabla V(t, \omega) = {}& \lambda \int_{\mathbb{R}^{2n}} \nabla^\top X(t, \omega) \nabla\phi(X(t, \omega) - X(t, \omega')) \\
& \otimes (V(t, \omega') - V(t, \omega))\,d\mu_0(\omega') \\
& - \lambda \nabla V(t, \omega) \int_{\mathbb{R}^{2n}} \phi(X(t, \omega) - X(t, \omega'))\,d\mu_0(\omega').
\end{aligned} \tag{4.21}$$

Thus,

$$\frac{d}{dt}\|\nabla X\|_{L^\infty(\Omega)} \le \|\nabla V\|_{L^\infty(\Omega)}, \tag{4.22}$$

and again with the use of a functional which maximizes $\ell(\nabla V(t, \omega))$, $\omega \in \Omega$,

$$\frac{\mathrm{d}}{\mathrm{d}t}\|\nabla V\|_{L^\infty(\Omega)} \le \lambda\|\nabla\phi\|_\infty M\|\nabla X\|_{L^\infty(\Omega)}\mathcal{A}_\Omega(t)$$

$$- \lambda M\phi(\mathcal{D}_\Omega(t))\|\nabla V\|_{L^\infty(\Omega)}. \tag{4.23}$$

For general kernels, we simply know the bound (4.17), and so the above calculation implies an exponential bound:

$$\|\nabla V\|_{L^\infty(\Omega)} + \|\nabla X\|_{L^\infty(\Omega)} \le C_1 e^{C_2 t}. \tag{4.24}$$

For heavy tail kernels, with the use of (4.20), we obtain a system of type similar to (2.30):

$$\frac{\mathrm{d}}{\mathrm{d}t}\|\nabla X\|_{L^\infty(\Omega)} \le \|\nabla V\|_{L^\infty(\Omega)}$$

$$\frac{\mathrm{d}}{\mathrm{d}t}\|\nabla V\|_{L^\infty(\Omega)} \le \lambda\|\nabla\phi\|_\infty M\|\nabla X\|_{L^\infty(\Omega)}\mathcal{A}_\Omega(0)e^{-t\lambda M\phi(\overline{\mathcal{D}}_\Omega)}$$

$$- \lambda M\phi(\overline{\mathcal{D}}_\Omega)\|\nabla V\|_{L^\infty(\Omega)}.$$

The resulting estimate (2.31) implies, noting that initially $\nabla(X_0, V_0) = \mathbb{I}$,

$$a\|\nabla X(t)\|^2_{L^\infty(\Omega)} + e^{bt}\|\nabla V(t)\|^2_{L^\infty(\Omega)} \le \frac{4\sqrt{a}}{b}(a+1), \tag{4.25}$$

where $a = \lambda M\|\nabla\phi\|_\infty\mathcal{A}_\Omega(0)$, $b = \lambda M\phi(\overline{\mathcal{D}}_\Omega)$. In particular, $\|\nabla V(t)\|_{L^\infty(\Omega)}$ is exponentially decaying.

4.4 Stability, Kantorovich-Rubinstein Metric, and Contractivity

In this section, we provide an analogue of the stability property of solutions to (4.1) similar to the one obtained in Sect. 4.4. On the kinetic level of description, the analytically suitable way to measure closeness of two flocks is via the use of Kantorovich-Rubinstein metric (or of Wasserstein-1 distance), which is compatible with the weak topology on $\mathcal{M}_+$. For two measures of equal mass $\mu, \nu \in \mathcal{M}^1_+$, we define

$$W_1(\mu, \nu) = \sup_{\mathrm{Lip}(g)\le 1}\left|\int_{\mathbb{R}^{2n}} g(\omega)\,\mathrm{d}\mu(\omega) - \int_{\mathbb{R}^{2n}} g(\omega)\,\mathrm{d}\nu(\omega)\right|. \tag{4.26}$$

It follows directly from the definition that if $\mathrm{Lip}(g) \le L$, then

$$\left| \int_{\mathbb{R}^{2n}} g(\omega) \, \mathrm{d}\mu(\omega) - \int_{\mathbb{R}^{2n}} g(\omega) \, \mathrm{d}\nu(\omega) \right| \leq L \, W_1(\mu, \nu).$$

Lemma 4.2 *For a sequence of measures with* $\mathrm{Supp}\,\mu_n \subset B_R(0)$, $W_1(\mu_n, \mu) \to 0$ *if and only if* $\mu_n \to \mu$ *weakly.*

Proof Clearly, if $W_1(\mu_n, \mu) \to 0$, then $\int_{\mathbb{R}^{2n}} g(\omega) \, \mathrm{d}\mu_n(\omega) \to \int_{\mathbb{R}^{2n}} g(\omega) \, \mathrm{d}\mu(\omega)$ for every Lipschitz function g. However, Lipschitz functions are dense in $C(B_{2R}(0))$; hence, $\mu_n \to \mu$ weakly. On the other hand, if $\mu_n \to \mu$ weakly, then $\mu_n \to \mu$ uniformly on any precompact subset of $C(B_{2R}(0))$, which, in particular, is the set of all g with $\mathrm{Lip}(g) \leq 1$. $\square$

So let us consider two solutions on a common interval of existence $\mu, \nu \in C_{w*}([0, T]; \mathcal{M}_+)$ with compact initial supports which we confine into a fixed compact domain Ω:

$$\mathrm{Supp}\,\mu_0 \cup \mathrm{Supp}\,\nu_0 \subset \Omega.$$

We also assume that the solutions have equal masses $M_\mu = M_\nu$ and momenta $\overline{V}_\mu = \overline{V}_\nu$. Clearly,

$$\frac{\mathrm{d}}{\mathrm{d}t} \|X_\mu(t) - X_\nu(t)\|_{L^\infty(\Omega)} \leq \|V_\mu(t) - V_\nu(t)\|_{L^\infty(\Omega)}.$$

For the velocities, we apply the same strategy as usual by fixing a maximizing functional ℓ and computing

$$\frac{\mathrm{d}}{\mathrm{d}t} \|V_\mu(t) - V_\nu(t)\|_{L^\infty(\Omega)} \leq \lambda \int_{\mathbb{R}^{2n}} \phi(X_\mu - X_\mu')\ell(V_\mu' - V_\mu) \, \mathrm{d}\mu_0(\omega')$$

$$- \lambda \int_{\mathbb{R}^{2n}} \phi(X_\nu - X_\nu')\ell(V_\nu' - V_\nu) \, \mathrm{d}\nu_0(\omega')$$

$$= \lambda \int_{\mathbb{R}^{2n}} \phi(X_\mu - X_\mu')\ell(V_\mu' - V_\mu)[\, \mathrm{d}\mu_0(\omega') - \mathrm{d}\nu_0(\omega')]$$

$$+ \lambda \int_{\mathbb{R}^{2n}} [\phi(X_\mu - X_\mu') - \phi(X_\nu - X_\nu')]\ell(V_\mu' - V_\mu) \, \mathrm{d}\nu_0(\omega')$$

$$+ \lambda \int_{\mathbb{R}^{2n}} \phi(X_\nu - X_\nu')\ell((V_\mu' - V_\nu') - (V_\mu - V_\nu)) \, \mathrm{d}\nu_0(\omega').$$

There are three terms to estimate on the right-hand side. For the first, we use the KR-distance:

$$\lambda \|\phi\|_{W^{1,\infty}} \left(\|\nabla X_\mu\|_{L^\infty(\Omega)} \mathcal{A}_{\mu,\Omega}(t) + \|\nabla V_\mu\|_{L^\infty(\Omega)} \right) W_1(\mu_0, \nu_0).$$

The second term in bounded by

$$2\lambda\|\nabla\phi\|_\infty M\|X_\mu(t) - X_\nu(t)\|_{L^\infty(\Omega)}\mathcal{A}_{\mu,\Omega}(t).$$

For the last term, we use maximality of $\ell(V_\mu - V_\nu)$ and pull out the kernel first:

$$\lambda\int_{\mathbb{R}^{2n}}\phi(X_\nu - X_\nu')\ell((V_\mu' - V_\nu') - (V_\mu - V_\nu))\,d\nu_0(\omega')$$

$$\leq \lambda\phi(\mathcal{D}_{\nu,\Omega})\int_{\mathbb{R}^{2n}}\ell((V_\mu' - V_\nu') - (V_\mu - V_\nu))\,d\nu_0(\omega')$$

$$= \lambda\phi(\mathcal{D}_{\nu,\Omega})\ell\left[\int_{\mathbb{R}^{2n}}(V_\mu' - V_\nu')\,d\nu_0(\omega')\right] - \lambda\phi(\mathcal{D}_{\nu,\Omega})M\|V_\mu(t) - V_\nu(t)\|_{L^\infty(\Omega)}$$

$$= \lambda\phi(\mathcal{D}_{\nu,\Omega})\ell\left[\int_{\mathbb{R}^{2n}}V_\mu'\,d\nu_0(\omega') - \overline{V}_\nu\right] - \lambda\phi(\mathcal{D}_{\nu,\Omega})M\|V_\mu(t) - V_\nu(t)\|_{L^\infty(\Omega)}$$

$$= \lambda\phi(\mathcal{D}_{\nu,\Omega})\ell\left[\int_{\mathbb{R}^{2n}}V_\mu'\,d\nu_0(\omega') - \int_{\mathbb{R}^{2n}}V_\mu'\,d\mu_0(\omega')\right]$$

$$- \lambda\phi(\mathcal{D}_{\nu,\Omega})M\|V_\mu(t) - V_\nu(t)\|_{L^\infty(\Omega)},$$

where in the last step we used equality of momenta. Continuing, we obtain

$$\leq \lambda\|\phi\|_\infty\|\nabla V_\mu\|_{L^\infty(\Omega)}W_1(\mu_0, \nu_0) - \lambda\phi(\mathcal{D}_{\nu,\Omega})M\|V_\mu(t) - V_\nu(t)\|_{L^\infty(\Omega)}.$$

Putting all the estimates together, we obtain the system:

$$\frac{d}{dt}\|X_\mu(t) - X_\nu(t)\|_{L^\infty(\Omega)} \leq \|V_\mu(t) - V_\nu(t)\|_{L^\infty(\Omega)}$$

$$\frac{d}{dt}\|V_\mu(t) - V_\nu(t)\|_{L^\infty(\Omega)} \leq \lambda\|\phi\|_{W^{1,\infty}}(\|\nabla X_\mu\|_{L^\infty(\Omega)}\mathcal{A}_{\mu,\Omega}(t)$$

$$+ \|\nabla V_\mu\|_{L^\infty(\Omega)})W_1(\mu_0, \nu_0)$$

$$+ 2\lambda\|\nabla\phi\|_\infty M\|X_\mu(t) - X_\nu(t)\|_{L^\infty(\Omega)}\mathcal{A}_{\mu,\Omega}(t)$$

$$- \lambda\phi(\mathcal{D}_{\nu,\Omega})M\|V_\mu(t) - V_\nu(t)\|_{L^\infty(\Omega)}.$$

Two conclusions follow as before from the system above. For general kernels ϕ, we can use the bound (4.24) to estimate

$$\frac{d}{dt}(\|X_\mu(t) - X_\nu(t)\|_{L^\infty(\Omega)} + \|V_\mu(t) - V_\nu(t)\|_{L^\infty(\Omega)})$$

$$\leq C_1 e^{C_2 t}W_1(\mu_0, \nu_0) + C_3\|X_\mu(t) - X_\nu(t)\|_{L^\infty(\Omega)}.$$

Since initially $\|X_\mu(0) - X_\nu(0)\|_{L^\infty(\Omega)} + \|V_\mu(0) - V_\nu(0)\|_{L^\infty(\Omega)} = 0$, we obtain

$$\|X_\mu(t) - X_\nu(t)\|_{L^\infty(\Omega)} + \|V_\mu(t) - V_\nu(t)\|_{L^\infty(\Omega)} \leq Ce^{Ct}W_1(\mu_0, \nu_0), \tag{4.27}$$

for some $C > 0$ which depends only on the initial condition and the kernel.

For heavy tail kernels, we use a more robust estimate on the deformation tensor (4.25) and on the diameter and amplitude (4.20) to conclude

$$\frac{\mathrm{d}}{\mathrm{d}t}\|V_\mu(t) - V_\nu(t)\|_{L^\infty(\Omega)} \leq ae^{-bt}[W_1(\mu_0, \nu_0) + \|X_\mu(t) - X_\nu(t)\|_{L^\infty(\Omega)}]$$
$$- b\|V_\mu(t) - V_\nu(t)\|_{L^\infty(\Omega)}.$$

So we obtain the same system (2.30) but for the new pair:

$$x = W_1(\mu_0, \nu_0) + \|X_\mu(t) - X_\nu(t)\|_{L^\infty(\Omega)}, \quad v = \|V_\mu(t) - V_\nu(t)\|_{L^\infty(\Omega)}.$$

Noting that $x(0) = W_1(\mu_0, \nu_0)$ and $v(0) = 0$, we obtain from (2.31)

$$\|X_\mu(t) - X_\nu(t)\|_{L^\infty(\Omega)} \leq CW_1(\mu_0, \nu_0),$$
$$\|V_\mu(t) - V_\nu(t)\|_{L^\infty(\Omega)} \leq Ce^{-ct}W_1(\mu_0, \nu_0), \tag{4.28}$$

for all time $t > 0$, where $C, c > 0$ depend only on the initial kinetic diameter of the flocks, mass, and the kernel.

Although (4.27) and (4.28) by themselves express characteristic stability of the flock, the ultimate application lies in estimating the KR-distance $W_1(\mu_t, \nu_t)$ and establishing contractivity of the kinetic dynamics. So let us assume that on a given time interval $[0, T)$, we have two solutions $\mu, \nu \in C_w([0, T); \mathcal{M}_+)$ with bounded supports Supp $\mu_0 \cup$ Supp $\nu_0 \subset \Omega$, where Ω is compact. Let us fix a function h with Lip$(h) \leq 1$ and use the conservation law (4.13):

$$\int_{\mathbb{R}^{2n}} h(\omega)\,\mathrm{d}\mu_t - \int_{\mathbb{R}^{2n}} h(\omega)\,\mathrm{d}\nu_t = \int_{\mathbb{R}^{2n}} h(X_\mu, V_\mu)\,\mathrm{d}\mu_0 - \int_{\mathbb{R}^{2n}} h(X_\nu, V_\nu)\,\mathrm{d}\nu_0$$
$$= \int_{\mathbb{R}^{2n}} h(X_\mu, V_\mu)(\mathrm{d}\mu_0 - \mathrm{d}\nu_0) + \int_{\mathbb{R}^{2n}} [h(X_\mu, V_\mu) - h(X_\nu, V_\nu)]\,\mathrm{d}\nu_0$$
$$\leq \mathrm{Lip}_\Omega(h(X_\mu, V_\mu))W_1(\mu_0, \nu_0)$$
$$+ M(\|X_\mu - X_\nu\|_{L^\infty(\Omega)} + \|V_\mu - V_\nu\|_{L^\infty(\Omega)}).$$

Using

$$\mathrm{Lip}_\Omega(h(X_\mu, V_\mu)) \leq \|\nabla V_\mu\|_{L^\infty(\Omega)} + \|\nabla X_\mu\|_{L^\infty(\Omega)}$$

and applying the stability and deformation estimates (4.24), (4.25), (4.27), and (4.28), we conclude the following bounds:

$$W_1(\mu_t, \nu_t) \leq Ce^{Ct}W_1(\mu_0, \nu_0), \quad \text{general kernels}, \tag{4.29}$$
$$W_1(\mu_t, \nu_t) \leq CW_1(\mu_0, \nu_0), \quad \text{heavy tail kernels}. \tag{4.30}$$

4.5   Mean-Field Limit

The mean-field limit refers to a passage from the discrete to the kinetic system as $N \to \infty$ in the scaling regime where the range of the interactions remains independent of N, i.e., $\phi(x, y)$ remains unrescaled.

The analysis of the previous section makes passing to the limit $N \to 0$ immediate. Let us start with a given measure $\mu_0 \in \mathcal{M}_+$ with compact support. We discretize it in the classical atomic approximation. We consider a box Q of side length L containing $\mathrm{Supp}\,\mu_0$, decompose it into N^n smaller boxes of side length L/N, and denote them Q_k, $k = 1, \ldots, N^n$ (see Fig. 4.1). Next we find a $\omega_k = (v_k, x_k)$ such that

$$v_k = \frac{1}{\mu_0(Q_k)} \int_{Q_k} v \, d\mu_0(x, v).$$

We dismiss those Q_k which have no mass. Finally, we define

$$\mu_0^N = \sum_{k=1}^{N^n} \mu_0(Q_k) \delta_{\omega_k}.$$

Clearly, all μ_0^N's have the same momentum and mass. Moreover, $\mu_0^N \to \mu_0$ weakly, and all supports $\mathrm{Supp}\,\mu_0^N$ are confined to the same box Q. Let us define the empirical measure:

$$\mu_t^N = \sum_{k=1}^{N^n} \mu_0(Q_k) \delta_{\omega_k(t)},$$

Fig. 4.1 Box decomposition of the support of μ_0.

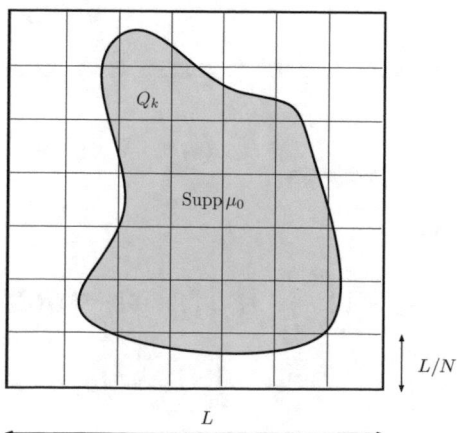

where $\omega_k(t) = (x_k(t), v_k(t))$ solves the Cucker-Smale system (2.1) with masses $m_k = \mu_0(Q_k)$. These measures will have uniformly bounded supports on any given time interval $[0, T]$ and in fact on all $\mathbb{R}^+$ if ϕ has heavy tail. Hence, the estimates (4.29) and (4.30) apply to give us

$$W_1(\mu_t^N, \mu_t^M) \le C(T) W_1(\mu_0^N, \mu_0^M), \quad \forall t < T,$$

and

$$W_1(\mu_t^N, \mu_t^M) \le C W_1(\mu_0^N, \mu_0^M), \quad \forall t > 0,$$

respectively. This means that at any time t, μ_t^N is a Cauchy sequence in $\mathcal{M}_+$; hence, there exists a weak limit $\mu_t^N \to \mu_t$, uniform on any $[0, T]$ and $\mathbb{R}^+$, respectively. The following lemma concludes the passage.

Lemma 4.3 (Stability Under Weak Limits) *Suppose a sequence of solutions* $\mu^n \in C_{w^*}([0, T); \mathcal{M}_+)$ *with* $\text{Supp}\,\mu_t^n \subset B_R(0)$ *for all* $t < T$ *and* $n \in \mathbb{N}$ *converges weakly pointwise, i.e.,* $\mu_t^n \to \mu_t$ *for all* $0 \le t < T$. *Then* $\mu \in C_{w^*}([0, T); \mathcal{M}_+)$ *is a weak solution to* (4.1).

Proof Weak continuity will follow immediately from (4.5) once it is established. It is clear that all the linear terms in (4.5) converge to the natural limits. As to $F(\mu_t^n)$, note that the family of functions $\{\phi(x-\cdot)(\cdot-v)\}_{(x,v)\in B_R(0)}$ is uniformly Lipschitz on $B_R(0)$, hence precompact in $C(B_R(0))$. So $F(\mu_t^n)(x, v) \to F(\mu_t)(x, v)$ converges uniformly on $B_R(0)$. This implies

$$\int_0^t \int_{\mathbb{R}^{2n}} F(\mu_s^n)(x, v) \cdot \nabla_v g(x, v) \, d\mu_s^n(x, v) \, ds \to$$

$$\to \int_0^t \int_{\mathbb{R}^{2n}} F(\mu_s)(x, v) \cdot \nabla_v g(x, v) \, d\mu_s(x, v) \, ds.$$

Indeed, adding and subtracting cross-terms, we obtain trivially

$$\int_0^t \int_{\mathbb{R}^{2n}} F(\mu_s)(x, v) \cdot \nabla_v g(x, v)(d\mu_s^n(x, v) - d\mu_s(x, v)) \, ds \to 0,$$

and

$$\left| \int_0^t \int_{\mathbb{R}^{2n}} [F(\mu_s^n)(x, v) - F(\mu_s)(x, v)] \cdot \nabla_v g(x, v) \, d\mu_s^n(x, v) \, ds \right|$$

$$\le \|\nabla g\|_\infty \int_0^t \int_{\mathbb{R}^{2n}} \|F(\mu_s^n) - F(\mu_s)\|_{L^\infty(B_R(0))} \, d\mu_s^n(x, v) \, ds$$

$$= C M_0^n \int_0^t \|F(\mu_s^n) - F(\mu_s)\|_{L^\infty(B_R(0))} \, ds \to 0.$$

□

If the initial measure μ_0 is in fact absolutely continuous $d\mu_0(\omega) = f_0(\omega) d\omega$, then the resulting solution will give density $f(t, \omega)$ obtained by the transport along characteristic flow according to (4.13). Explicitly, we have the Liouville formula for the Jacobian of the flow:

$$\det \nabla_\omega (X, V)(\omega, t) = \exp\left\{ -\lambda n \int_0^t \phi * \rho(X(\omega, s), s) \, ds \right\},$$

where ρ is the macroscopic density measure $d\rho(x, t) = \int_{\mathbb{R}^n} \mu_t(x, v)$, i.e., first marginal of μ_t. Note that $\phi * \rho \in C^\infty$. Then

$$f(X(\omega, t), V(\omega, t), t) = f_0(\omega) \exp\left\{ \lambda n \int_0^t \phi * \rho(X(\omega, s), s) \, ds \right\}.$$

Inverting the flow, we recover $f(t)$. It is clear that $f(t)$ inherits smoothness of the initial condition as well, in view of the smoothness of the flow map. At the C^1 level, it is seen from (4.24). Further regularity can be deduced by proving higher regularity of the flow map.

4.6   Notes and References

The kinetic alignment model was first derived in Ha and Tadmor [45] and shortly after justified through the mean-field limit by Ha and Liu [43]. Asymptotic flocking based on the characteristic flow approach was proved in Carrillo et al. [13], where other Boltzmann-type kinetic models were derived; see [1, 14] for comprehensive surveys. Stability estimates were established in [49]. A kinetic version of the control problem considered in [12] was addressed by Piccoli, Rossi, and Trélat in [84]. The main result is existence of a control mechanism with finite time support which steers the kinetic transport into alignment. The corrector method, technically, has never been applied at the kinetic level. However, it is likely to lead to a similar result; see Chap. 6 for the hydrodynamic version.

   Justification of kinetic formulation for singular models remains a challenging problem even at the stage of defining the field $F(\mu)$ unless singularity is integrable. The difficulty is more manageable in the weakly singular case $\beta < \frac{1}{2}$, where Mucha and Peszek [77] established the mean-field limit and existence results.

Chapter 5
Macroscopic Description: Hydrodynamic Limit

By taking v-moments of (4.1), we can read off the system for macroscopic density and momentum:

$$\begin{cases} \rho_t + \nabla \cdot (\rho \mathbf{u}) = 0, \\ (\rho \mathbf{u})_t + \nabla_x \cdot (\rho \mathbf{u} \otimes \mathbf{u} + \mathcal{R}) = \displaystyle\int_{\mathbb{R}^n} \rho(x)\rho(y)(\mathbf{u}(y) - \mathbf{u}(x))\phi(x - y)\, dy, \end{cases}$$

where $\mathcal{R}$ is the Reynolds stress tensor:

$$\mathcal{R}(x, t) = \int_{\mathbb{R}^n} (v - \mathbf{u}(x, t)) \otimes (v - \mathbf{u}(x, t)) f(x, v, t)\, dv.$$

One can formally close the system by considering a monokinetic density ansatz concentrated at the macroscopic velocity $\mathbf{u}$:

$$f(x, v, t) = \rho(x, t)\delta(v - \mathbf{u}(x, t)). \tag{5.1}$$

Clearly, such an ansatz removes the stress, $\mathcal{R} = 0$, and hence, we obtain a closed system:

$$\begin{cases} \rho_t + \nabla \cdot (\rho \mathbf{u}) = 0, \\ (\rho \mathbf{u})_t + \nabla_x \cdot (\rho \mathbf{u} \otimes \mathbf{u}) = \displaystyle\int_{\mathbb{R}^n} \rho(x)\rho(y)(\mathbf{u}(y) - \mathbf{u}(x))\phi(x - y)\, dy. \end{cases} \tag{5.2}$$

One drawback of dealing with the system in conservation form is that the momentum variable $\rho \mathbf{u}$ does not enjoy the maximum principle, inherent for the alignment dynamics. Moreover, from the point of view of studying regularity, writing (5.2) for the momentum-density pair would require division by ρ: $\rho \mathbf{u} \otimes \mathbf{u} = \frac{1}{\rho}(\rho \mathbf{u}) \otimes (\rho \mathbf{u})$,

© Springer Nature Switzerland AG 2021

R. Shvydkoy, *Dynamics and Analysis of Alignment Models of Collective Behavior*, Nečas Center Series, https://doi.org/10.1007/978-3-030-68147-0_5

which necessitates no-vacuum assumption on solutions. Instead we will study the system for velocity-density pair:

$$
\begin{cases}
\rho_t + \nabla \cdot (\rho \mathbf{u}) = 0, \\[2mm]
\mathbf{u}_t + \mathbf{u} \cdot \nabla \mathbf{u} = \displaystyle\int_{\mathbb{R}^n} \rho(y)(\mathbf{u}(y) - \mathbf{u}(x))\phi(x - y)\, dy.
\end{cases}
\tag{5.3}
$$

It is called the *pressureless Euler alignment system*, EAS for short. Clearly, (5.2) and (5.3) are equivalent under no-vacuum condition, and solutions of (5.3) are also solutions of (5.2) for any pair $(\rho, \mathbf{u})$ with or without vacuum. Moreover, EAS (5.3) enjoys maximum principle and is more amenable to well-posedness analysis, which we will discuss in great detail in coming chapters.

The goal of this chapter is to provide a rigorous derivation of the pressureless Euler alignment system from the kinetic description. Let us note right away that for a smooth pair $(\rho, \mathbf{u})$, the monokinetic ansatz (5.1) is a measure-valued solution to the kinetic model (4.1) in the sense of Definition 4.1 if and only if it is a solution to the EAS (5.3). However, it is not a satisfactory justification considering that *smooth* solutions are more practical in use. One way to force a smooth distribution f to converge to the monokinetic one is to consider a modified model where solutions are being penalized for deviating from the monokinetic measure. To achieve this, we consider a model with strong local alignment:

$$
\partial_t f^\varepsilon + v \cdot \nabla_x f^\varepsilon + \lambda \nabla_v \cdot [f^\varepsilon F(f^\varepsilon)] + \frac{1}{\varepsilon}\nabla_v[f^\varepsilon(\mathbf{u}^\varepsilon_{\rho^\varepsilon,\delta} - v)] = 0,
\tag{5.4}
$$

where $\mathbf{u}^\varepsilon_{\rho^\varepsilon,\delta}$ is a special density-weighted Favre filtration to be defined precisely in the next section. We will justify mean-field limit from the corresponding discrete system for this mollified model before we move to the main goal of this chapter, which is passage from kinetic (5.4) to pressureless EAS. Namely, we show in Theorem 5.1 that all solutions to (5.4) converge to the monokinetic one (5.1) as long as $(\rho, \mathbf{u})$ solves the EAS (5.3) in the scaling regime:

$$
\delta \sim \varepsilon^2.
$$

The result will be proved on $\mathbb{R}^n$ with bounded support or on the torus.

5.1   Multi-Scale Model and Its Justification

Consider the mollifier

$$
\psi(r) = \frac{c}{(1 + r^2)^{\frac{n+\gamma}{2}}}, \qquad \gamma > 1,
\tag{5.5}
$$

where c is a normalizing constant so that $\int_{\mathbb{R}^n} \psi(|x|)\,dx = 1$. Denote $\psi_\delta = \frac{1}{\delta^n}\psi(r/\delta)$. We fix a density $\rho \in L^1(\mathbb{R}^n)$ in what follows. For a function $g \in L^\infty(\mathbb{R}^n)$, we consider the standard mollification $g_\delta = \psi_\delta * g$ and a special density-weighted mollification, called *Favre filtration*, defined by[1]

$$g_{\rho,\delta} = \left(\frac{(g\rho)_\delta}{\rho_\delta}\right)_\delta. \tag{5.6}$$

Let us note right away that $\rho_\delta(x) > 0$ for all $x \in \mathbb{R}^n$ due to infinite tails of the mollifier. At the same time,

$$\left\|\frac{(g\rho)_\delta}{\rho_\delta}\right\|_\infty \le \|g\|_\infty.$$

This insures that further mollification is legitimate. The resulting function is locally smooth with a few remarkable properties which we address next.

First, the Favre filtration is a symmetric operation relative to the ρ-weighted inner product:

$$\langle f, g\rangle_\rho = \int_{\mathbb{R}^n} f(x)g(x)\rho(x)\,dx. \tag{5.7}$$

We have

$$\langle f_{\rho,\delta}, g\rangle_\rho = \langle f, g_{\rho,\delta}\rangle_\rho. \tag{5.8}$$

Second, if $g \in W^{1,\infty}$, its Favre filtration approximates g well, and what is crucial, it does not rely on any regularity of the density itself, just its mass M. In the following lemma, we denote L^1_ρ the L^1 space over the density measure $\rho\,dx$.

Lemma 5.1 *Let $g \in W^{1,\infty}$. Then*

$$\|g_{\rho,\delta} - g\|_{L^1_\rho} \le \delta C_\psi M \|g\|_{W^{1,\infty}}. \tag{5.9}$$

Proof We have

$$\|g_{\rho,\delta} - g\|_{L^1_\rho} = \sup_{\|f\|_\infty \le 1} \int_{\mathbb{R}^n} f(g_{\rho,\delta} - g)\rho\,dx.$$

So let us fix $\|f\|_\infty \le 1$ and compute

---

[1]Technically, the original Favre filtration is defined by $\frac{(g\rho)_\delta}{\rho_\delta}$, [40], but we still adapt the same term for our over-mollified version.

$$\int_{\mathbb{R}^n} f(g_{\rho,\delta} - g)\rho \, \mathrm{d}x = \int_{\mathbb{R}^n} f(g_{\rho,\delta} - g_\delta)\rho \, \mathrm{d}x + \int_{\mathbb{R}^n} f(g_\delta - g)\rho \, \mathrm{d}x.$$

In the last integral, we simply use $\|g_\delta - g\|_\infty \le C\delta\|g\|_{W^{1,\infty}}$. So

$$\left| \int_{\mathbb{R}^n} f(g_\delta - g)\rho \, \mathrm{d}x \right| \le CM\delta\|g\|_{W^{1,\infty}}.$$

In the first integral, we switch the outer mollification back on $f\rho$:

$$\int_{\mathbb{R}^n} f(g_{\rho,\delta} - g_\delta)\rho \, \mathrm{d}x = \int_{\mathbb{R}^n} (f\rho)_\delta \left(\frac{(g\rho)_\delta}{\rho_\delta} - g \right) \mathrm{d}x$$

$$= \int_{\mathbb{R}^n} \frac{(f\rho)_\delta}{\rho_\delta} ((g\rho)_\delta - g\rho_\delta) \, \mathrm{d}x.$$

We have

$$(g\rho)_\delta(x) - g(x)\rho_\delta(x) = \int_{\mathbb{R}^n} \psi_\delta(x - y)\rho(y)[g(y) - g(x)] \, \mathrm{d}y$$

$$\le \delta\|g\|_{W^{1,\infty}} \int_{\mathbb{R}^n} \tilde{\psi}_\delta(x - y)\rho(y) \, \mathrm{d}y,$$

where $\tilde{\psi}(r) = r\psi(r)$, still an integrable kernel. At the same time,

$$\frac{(f\rho)_\delta(x)}{\rho_\delta(x)} \le \|f\|_\infty \le 1, \qquad \forall x \in \mathbb{R}^n.$$

So

$$\left| \int_{\mathbb{R}^n} f(g_{\rho,\delta} - g_\delta)\rho \, \mathrm{d}x \right| \le \delta\|g\|_{W^{1,\infty}} \int_{\mathbb{R}^{2n}} \tilde{\psi}_\delta(x - y)\rho(y) \, \mathrm{d}y \, \mathrm{d}x \le CM\delta\|g\|_{W^{1,\infty}}.$$

$$\square$$

For a fixed pair of parameters $\varepsilon, \delta > 0$, we consider the model given by

$$\partial_t f^\varepsilon + v \cdot \nabla_x f^\varepsilon + \lambda \nabla_v \cdot [f^\varepsilon F(f^\varepsilon)] + \frac{1}{\varepsilon} \nabla_v [f^\varepsilon(\mathbf{u}^{\varepsilon,\delta} - v)] = 0, \tag{5.10}$$

$$\mathbf{u}^{\varepsilon,\delta} := \mathbf{u}^\varepsilon_{\rho^\varepsilon,\delta}.$$

Notably, $\mathbf{u}^{\varepsilon,\delta}$ is defined only in terms of the macroscopic momentum $\rho\mathbf{u}$ and density, which are marginals of f. So even if $f = \mu$ is a measure, the mollifications $(\rho\mathbf{u})_\delta$, ρ_δ are smooth functions, and as stated earlier, $\rho_\delta > 0$. Moreover, if μ has a bounded support, then

$$\frac{(\rho \mathbf{u})_\delta}{\rho_\delta}(x) = \frac{\int_{\mathbb{R}^{2n}} \psi_\delta(x-y)w\,d\mu(y,w)}{\int_{\mathbb{R}^{2n}} \psi_\delta(x-y)\,d\mu(y,w)} \leq \mathrm{diam}(\mathrm{Supp}\,\mu).$$

This defines a bounded function, which then produces the smooth field $\mathbf{u}^{\varepsilon,\delta}$ we used to define the model. This makes the model extendable to the class of measure-valued solutions as in Definition 4.1. In particular, one can verify directly that the empirical measure (4.3) solves (5.10) if and only if the discrete variables $(\mathbf{x}_i, \mathbf{v}_i)$ satisfy

$$\begin{cases} \dot{\mathbf{x}}_i = \mathbf{v}_i, \\[2mm] \dot{\mathbf{v}}_i = \lambda \sum_{j=1}^{N} m_j \phi(\mathbf{x}_i - \mathbf{x}_j)(\mathbf{v}_j - \mathbf{v}_i) \\[2mm] \quad + \frac{1}{\varepsilon} \int_{\mathbb{R}^n} \psi_\delta(\mathbf{x}_i - y) \frac{\sum_k m_k \psi_\delta(\mathbf{x}_k - y)(\mathbf{v}_k - \mathbf{v}_i)}{\sum_k m_k \psi_\delta(\mathbf{x}_k - y)}\,dy, \end{cases} \qquad (5.11)$$

Note that this represents an averaged version of the Motsch-Tadmor model which has a restored symmetry—its total momentum is conserved. The same is true of the kinetic model (5.10). Clearly, the system (5.11) is globally well-posed.

We now address the mean-field limit procedure. Following the methodology carried out in the classical case, we only need to assess the contribution of the new term in (5.10). The corresponding V-characteristics read

$$\frac{d}{dt}V(\omega,t) = \lambda F(\mu_t)(X,V) + \frac{1}{\varepsilon}G_\delta(X,V),$$

where

$$G_\delta = \int_{\mathbb{R}^n} \psi_\delta(X-y) \frac{\int_{\mathbb{R}^{2n}} \psi_\delta(y - X(\omega',t))[V(\omega',t) - V(\omega,t)]\,d\mu_0(\omega')}{\int_{\mathbb{R}^{2n}} \psi_\delta(y - X(\omega',t))\,d\mu_0(\omega')}\,dy.$$

So we have the maximum principle, and consequently, the results of Sect. 4.3 on amplitude and alignment estimates apply. As far as deformation estimates are concerned, it is clearly not expected that they will remain independent of ε, δ. But so long as ε, δ are fixed, the extra terms we deal with still remain smooth. So we fix a finite time interval $[0, T]$ and a compact domain Ω containing $\mathrm{Supp}\,\mu_0$. With the use of a uniform bound on the amplitude $\mathcal{A}_\Omega$, we obtain, following our estimates in (4.21)–(4.23),

$$\frac{d}{dt}\|\nabla V\|_{L^\infty(\Omega)} \leq C\|\nabla X\|_{L^\infty(\Omega)} + \frac{1}{\varepsilon}\|\nabla G_\delta(X,V)\|_{L^\infty(\Omega)},$$

while differentiating G_δ, we obtain

$$\|\nabla G_\delta(X, V)\|_{L^\infty(\Omega)} \lesssim \frac{1}{\delta}\|\nabla X\|_{L^\infty(\Omega)} + \|\nabla V\|_{L^\infty(\Omega)}.$$

So for the quantity $Y = \|\nabla X\|_{L^\infty(\Omega)} + \|\nabla V\|_{L^\infty(\Omega)}$, we obtain Grönwall's lemma, $Y_t \leq C_{\varepsilon,\delta} Y$. Consequently, we have a bound on gradients on any finite time interval:

$$\|\nabla X(t)\|_{L^\infty(\Omega)} + \|\nabla V(t)\|_{L^\infty(\Omega)} \leq C(T, \delta, \varepsilon). \tag{5.12}$$

Moving on to stability estimates, we use the same setup as in Sect. 4.4. We find

$$\frac{d}{dt}\|V_\mu(t) - V_\nu(t)\|_{L^\infty(\Omega)} \leq C_T W_1(\mu_0, \nu_0) + C\|X_\mu(t) - X_\nu(t)\|_{L^\infty(\Omega)}$$

$$+ \frac{1}{\varepsilon}\|G_\delta(X_\mu, V_\mu)(t) - G_\delta(X_\nu, V_\nu)(t)\|_{L^\infty(\Omega)}.$$

It remains to estimate the last term. We denote for short $V' = V(\omega', t)$, $X' = X(\omega', t)$, etc. We obtain

$$G_\delta(X_\mu, V_\mu) - G_\delta(X_\nu, V_\nu) = \int_{\mathbb{R}^n} \psi_\delta(X_\mu - y)\frac{\int_{\mathbb{R}^{2n}} \psi_\delta(y - X'_\mu)[V'_\mu - V_\mu]\,d\mu_0}{\int_{\mathbb{R}^{2n}} \psi_\delta(y - X'_\mu)\,d\mu_0}\,dy$$

$$- \int_{\mathbb{R}^n} \psi_\delta(X_\nu - y)\frac{\int_{\mathbb{R}^{2n}} \psi_\delta(y - X'_\nu)[V'_\nu - V_\nu]\,d\nu_0}{\int_{\mathbb{R}^{2n}} \psi_\delta(y - X'_\nu)\,d\nu_0}\,dy.$$

Let us break it down further into

$$\sum_{i=1}^{5} J_i = \int_{\mathbb{R}^n} \psi_\delta(X_\mu - y)\frac{\int_{\mathbb{R}^{2n}} \psi_\delta(y - X'_\mu)[V'_\mu - V'_\nu + V_\nu - V_\mu]\,d\mu_0}{\int_{\mathbb{R}^{2n}} \psi_\delta(y - X'_\mu)\,d\mu_0}\,dy$$

$$+ \int_{\mathbb{R}^n} \psi_\delta(X_\mu - y)\frac{\int_{\mathbb{R}^{2n}} \psi_\delta(y - X'_\mu)[V'_\nu - V_\nu][\,d\mu_0 - d\nu_0]}{\int_{\mathbb{R}^{2n}} \psi_\delta(y - X'_\mu)\,d\mu_0}\,dy$$

$$+ \int_{\mathbb{R}^n} \psi_\delta(X_\mu - y)\frac{\int_{\mathbb{R}^{2n}} [\psi_\delta(y - X'_\mu) - \psi_\delta(y - X'_\nu)][V'_\nu - V_\nu]\,d\nu_0}{\int_{\mathbb{R}^{2n}} \psi_\delta(y - X'_\mu)\,d\mu_0}\,dy$$

$$+ \int_{\mathbb{R}^n} [\psi_\delta(X_\mu - y) - \psi_\delta(X_\nu - y)]\frac{\int_{\mathbb{R}^{2n}} \psi_\delta(y - X'_\nu)[V'_\nu - V_\nu]\,d\nu_0}{\int_{\mathbb{R}^{2n}} \psi_\delta(y - X'_\mu)\,d\mu_0}\,dy$$

$$+ \int_{\mathbb{R}^n} \psi_\delta(X_\nu - y)\int_{\mathbb{R}^{2n}} \psi_\delta(y - X'_\nu)[V'_\nu - V_\nu]\,d\nu_0\left[\frac{1}{\int_{\mathbb{R}^{2n}} \psi_\delta(y - X'_\mu)\,d\mu_0}\right.$$

$$\left. - \frac{1}{\int_{\mathbb{R}^{2n}} \psi_\delta(y - X'_\nu)\,d\nu_0}\right]\,dy.$$

Let us estimate term by term. The first one is straightforward:

$$\|J_1\|_{L^\infty(\Omega)} \leq C\|V_\mu(t) - V_\nu(t)\|_{L^\infty(\Omega)}.$$

In the remaining terms, we rely on the following observation. Since we work on a finite interval $t \leq T$, all the characteristics have a finite reach $\|X_\mu\|_{L^\infty(\Omega)} + \|X_\nu\|_{L^\infty(\Omega)} \leq R_T$. So if $|y| < 2R_T$, we have $C_\delta \geq \psi_\delta(y - X(\omega', t)) \geq c_\delta$, for all $\omega' \in \Omega$ and $t < T$. Otherwise, if $|y| \geq 2R_T$, we have directly from the definition of the mollifier (5.5)

$$c\psi_\delta(y) \leq \psi_\delta(y - X(\omega', t)) \leq C\psi_\delta(y),$$
$$c|\nabla\psi_\delta(y)| \leq |\nabla\psi_\delta(y - X(\omega', t))| \leq C|\nabla\psi_\delta(y)|. \tag{5.13}$$

With this in mind, we estimate the remaining terms $J_2, \ldots, J_5$ by splitting the region of integration into $|y| < 2R_T$ and $|y| \geq 2R_T$. Thus, for J_2, we have, for $|y| < 2R_T$, in view of (5.12),

$$\int_{\mathbb{R}^{2n}} \psi_\delta(y - X'_\mu)[V'_\nu - V_\nu][d\mu_0 - d\nu_0] \leq CW_1(\mu_0, \nu_0),$$

while for $|y| \geq 2R_T$,

$$\int_{\mathbb{R}^{2n}} \psi_\delta(y - X'_\mu)[V'_\nu - V_\nu][d\mu_0 - d\nu_0] \leq C_\delta(\psi_\delta(y) + |\nabla\psi_\delta(y)|)W_1(\mu_0, \nu_0).$$

Hence,

$$J_2 \lesssim W_1(\mu_0, \nu_0) \int_{|y| < 2R_T} \frac{\psi_\delta(X_\mu - y)}{\int_{\mathbb{R}^{2n}} \psi_\delta(y - X'_\mu)\,d\mu_0}\,dy$$
$$+ W_1(\mu_0, \nu_0) \int_{|y| \geq 2R_T} \frac{\psi_\delta(y) + |\nabla\psi_\delta(y)|}{\int_{\mathbb{R}^{2n}} \psi_\delta(y - X'_\mu)\,d\mu_0}\psi_\delta(X_\mu - y)\,dy.$$

The first integral is simply bounded by a constant, and hence the term $\lesssim W_1(\mu_0, \nu_0)$. In the second integral, in view of (5.13),

$$\int_{|y| \geq 2R_T} \frac{\psi_\delta(y) + |\nabla\psi_\delta(y)|}{\int_{\mathbb{R}^{2n}} \psi_\delta(y - X'_\mu)\,d\mu_0}\psi_\delta(X_\mu - y)\,dy$$
$$\sim \int_{|y| \geq 2R_T} \frac{\psi_\delta(y) + |\nabla\psi_\delta(y)|}{\int_{\mathbb{R}^{2n}} \psi_\delta(y)\,d\mu_0}\psi_\delta(y)\,dy$$
$$\sim \int_{|y| \geq 2R_T} \psi_\delta(y) + |\nabla\psi_\delta(y)|\,dy = C.$$

Thus,

$$J_2 \leq CW_1(\mu_0, \nu_0).$$

Proceeding the remaining terms, we argue similarly. For J_3, we have for $|y| < 2R_T$

$$|\psi_\delta(y - X'_\mu) - \psi_\delta(y - X'_\nu)| \leq \|X_\mu(t) - X_\nu(t)\|_{L^\infty(\Omega)},$$

and for $|y| \geq 2R_T$,

$$|\psi_\delta(y - X'_\mu) - \psi_\delta(y - X'_\nu)| \leq |\nabla\psi_\delta(y)|\|X_\mu(t) - X_\nu(t)\|_{L^\infty(\Omega)}.$$

Consequently, by the same computation as above,

$$J_3 \leq C\|X_\mu(t) - X_\nu(t)\|_{L^\infty(\Omega)}.$$

Similarly, for J_4, we have the exact same difference of kernels in the outer integral. So the same argument applies:

$$J_4 \leq C\|X_\mu(t) - X_\nu(t)\|_{L^\infty(\Omega)}.$$

And finally,

$$\frac{1}{\int_{\mathbb{R}^{2n}} \psi_\delta(y - X'_\mu)\,\mathrm{d}\mu_0} - \frac{1}{\int_{\mathbb{R}^{2n}} \psi_\delta(y - X'_\nu)\,\mathrm{d}\nu_0}$$

$$= \frac{\int_{\mathbb{R}^{2n}} \psi_\delta(y - X'_\nu)\,\mathrm{d}\nu_0 - \int_{\mathbb{R}^{2n}} \psi_\delta(y - X'_\mu)\,\mathrm{d}\mu_0}{\int_{\mathbb{R}^{2n}} \psi_\delta(y - X'_\mu)\,\mathrm{d}\mu_0 \int_{\mathbb{R}^{2n}} \psi_\delta(y - X'_\nu)\,\mathrm{d}\nu_0}$$

$$= \frac{\int_{\mathbb{R}^{2n}} [\psi_\delta(y - X'_\nu) - \psi_\delta(y - X'_\mu)]\,\mathrm{d}\nu_0 + \int_{\mathbb{R}^{2n}} \psi_\delta(y - X'_\mu)[\,\mathrm{d}\nu_0 - \mathrm{d}\mu_0]}{\int_{\mathbb{R}^{2n}} \psi_\delta(y - X'_\mu)\,\mathrm{d}\mu_0 \int_{\mathbb{R}^{2n}} \psi_\delta(y - X'_\nu)\,\mathrm{d}\nu_0}.$$

All the terms are similar to those we encountered before, so we obtain the same estimate:

$$J_5 \leq C\|X_\mu(t) - X_\nu(t)\|_{L^\infty(\Omega)} + CW_1(\mu_0, \nu_0).$$

Ultimately, we obtain

$$\|G_\delta(X_\mu, V_\mu)(t) - G_\delta(X_\nu, V_\nu)(t)\|_{L^\infty(\Omega)} \lesssim W_1(\mu_0, \nu_0) + \|X_\mu(t) - X_\nu(t)\|_{L^\infty(\Omega)}$$
$$+ \|V_\mu(t) - V_\nu(t)\|_{L^\infty(\Omega)}.$$

Then the grand quantity $Y = \|X_\mu(t) - X_\nu(t)\|_{L^\infty(\Omega)} + \|V_\mu(t) - V_\nu(t)\|_{L^\infty(\Omega)}$ satisfies

$$\frac{\mathrm{d}}{\mathrm{d}t} Y \lesssim W_1(\mu_0, \nu_0) + Y.$$

Given that $Y_0 = 0$, we obtain

$$Y(t) \lesssim W_1(\mu_0, \nu_0), \qquad t < T.$$

With the stability and deformation estimates at hand, we conclude as in Sect. 4.4:

$$W_1(\mu_t, \nu_t) \lesssim W_1(\mu_0, \nu_0), \qquad t < T.$$

This establishes convergence of the empirical measures toward a solution to (5.10). Moreover, if f_0^ε is a smooth distribution, it will remain so for all times.

5.2 Hydrodynamic Limit: Kinetic Relative Entropy

We work in the settings of a finite time interval $[0, T]$ and all initial conditions f_0^ε to (5.4) being bounded to a common ball $\{|\omega| \le R_0\}$. This guarantees, as we noted in the previous sections, that $\operatorname{Supp} f^\varepsilon(t) \subset \{|\omega| \lesssim R + t\}, t \le T$. So we have a common bound on supports of constructed solutions on a given time interval.

Theorem 5.1 *Let $(\rho, \mathbf{u})$ be a classical solution to (5.3) on the time interval $[0, T)$. Let $f = \rho(x, t)\delta_{v=\mathbf{u}(x,t)}$ be the corresponding monokinetic solution to (4.1). Suppose $f_0^\varepsilon \in C^1(\mathbb{R}^{2n})$ is a family of initial conditions for (5.4) satisfying*

(F1) $\operatorname{Supp} f_0^\varepsilon \subset \{|w| < R_0\}$;
(F2) $W_1(f_0^\varepsilon, f_0) \le \varepsilon$.

Then in the scaling regime $\delta = \varepsilon^2$, we have, for all $t < T$,

$$W_1(f^\varepsilon(t), f(t)) + W_1(\rho^\varepsilon(t), \rho(t)) \le C\sqrt{\varepsilon}.$$

Remark 5.1 The proof shows convergence for more general parameter δ as long as $\delta = o(\varepsilon)$. Specifically,

$$W_1(f^\varepsilon(t), f(t)) + W_1(\rho^\varepsilon(t), \rho(t)) \lesssim \sqrt{\varepsilon + \frac{\delta}{\varepsilon}}.$$

We see that $\delta = \varepsilon^2$ appears to be most optimal: if $\delta \ll \varepsilon^2$, the kinetic equation becomes over-resolved without improvement on convergence rate of solutions; if $\delta \gg \varepsilon^2$, the model is under-resolved and the convergence rate slows down.

Letting

$$\mathcal{E}^\varepsilon = \frac{1}{2} \int_{\mathbb{R}^{2n}} |v|^2 f^\varepsilon(x, v) \, dx \, dv,$$

$$\mathcal{I}_2^\varepsilon = \frac{1}{2} \int_{\mathbb{R}^{2n}} \int_{\mathbb{R}^{2n}} \phi(x - y)|w - v|^2 f^\varepsilon(y, w, t) f^\varepsilon(x, v, t) \, dy \, dw \, dx \, dv$$

we have the following energy balance relation:

$$\frac{d}{dt} \mathcal{E}^\varepsilon = -\mathcal{I}_2^\varepsilon + \frac{1}{\varepsilon} \int_{\mathbb{R}^n} \frac{|(\rho^\varepsilon \mathbf{u}^\varepsilon)_\delta|^2}{\rho_\delta^\varepsilon} \, dx - \frac{2}{\varepsilon} \mathcal{E}^\varepsilon. \qquad (5.14)$$

The key observation is that the local alignment term in the energy law controls inner energies both relative to the local field $\mathbf{u}^\varepsilon$ and to its Favre filtration $\mathbf{u}^{\varepsilon,\delta}$. Let us first define

$$\mathbf{u}^\varepsilon(x) = \begin{cases} \dfrac{\rho^\varepsilon \mathbf{u}^\varepsilon(x)}{\rho^\varepsilon(x)}, & \rho^\varepsilon(x) > 0, \\[3mm] 0, & \rho^\varepsilon(x) = 0. \end{cases}$$

This defines a bounded but possibly non-smooth extension of $\mathbf{u}^\varepsilon$ from the support of the density. However, in the course of the proof, we will not rely on any regularity of $\mathbf{u}^\varepsilon$ itself.

So we have the following formulas for internal energies:

$$\mathcal{E}_I^\varepsilon := \frac{1}{2} \int_{\mathbb{R}^{2n}} |v - \mathbf{u}^\varepsilon(x)|^2 f^\varepsilon(x, v) \, dx \, dv \quad = \mathcal{E}^\varepsilon - \frac{1}{2} \int_{\mathbb{R}^n} \rho^\varepsilon(x)|\mathbf{u}^\varepsilon(x)|^2 \, dx$$

$$\mathcal{E}_I^{\varepsilon,\delta} := \frac{1}{2} \int_{\mathbb{R}^{2n}} |v - \mathbf{u}^{\varepsilon,\delta}(x)|^2 f^\varepsilon(x, v) \, dx \, dv \quad = \mathcal{E}^\varepsilon - \int_{\mathbb{R}^n} \frac{|(\rho^\varepsilon \mathbf{u}^\varepsilon)_\delta|^2}{\rho_\delta^\varepsilon} \, dx$$

$$+ \frac{1}{2} \int_{\mathbb{R}^n} \rho^\varepsilon(x)|\mathbf{u}^{\varepsilon,\delta}(x)|^2 \, dx.$$

The two energies add up to

$$\mathcal{E}_I^\varepsilon + \mathcal{E}_I^{\varepsilon,\delta} = 2\mathcal{E}^\varepsilon - \int_{\mathbb{R}^n} \frac{|(\rho^\varepsilon \mathbf{u}^\varepsilon)_\delta|^2}{\rho_\delta^\varepsilon} \, dx + \frac{1}{2} \int_{\mathbb{R}^n} \rho^\varepsilon(x)|\mathbf{u}^{\varepsilon,\delta}(x)|^2 \, dx$$

$$- \frac{1}{2} \int_{\mathbb{R}^n} \rho^\varepsilon(x)|\mathbf{u}^\varepsilon(x)|^2 \, dx.$$

We now show that the last two terms add up to a negative value. Indeed, for any pair $(\mathbf{u}, \rho)$, we have by Minkowski's inequality

$$|\mathbf{u}_{\rho,\delta}|^2 \leq \left(\left| \frac{(\rho \mathbf{u})_\delta}{\rho_\delta} \right|^2 \right)_\delta \leq \left(\frac{(\rho |\mathbf{u}|^2)_\delta}{\rho_\delta} \right)_\delta .$$

Thus,

$$\int_{\mathbb{R}^n} \rho(x) |\mathbf{u}_{\rho,\delta}(x)|^2 \, dx \leq \int_{\mathbb{R}^n} \rho(x) \left(\frac{(\rho |\mathbf{u}|^2)_\delta}{\rho_\delta} \right)_\delta (x) \, dx$$

$$= \int_{\mathbb{R}^n} \rho_\delta(x) \frac{(\rho |\mathbf{u}|^2)_\delta}{\rho_\delta}(x) \, dx = \int_{\mathbb{R}^n} (\rho |\mathbf{u}|^2)_\delta(x) \, dx = \int_{\mathbb{R}^n} \rho(x) |\mathbf{u}(x)|^2 \, dx.$$

We have shown that

$$2\mathcal{E}^\varepsilon - \int_{\mathbb{R}^n} \frac{|(\rho^\varepsilon \mathbf{u}^\varepsilon)_\delta|^2}{\rho_\delta^\varepsilon} \, dx \geq \mathcal{E}_I^\varepsilon + \mathcal{E}_I^{\varepsilon,\delta}.$$

Consequently, the following energy inequality holds:

$$\frac{d}{dt} \mathcal{E}^\varepsilon \leq -\mathcal{I}_2^\varepsilon - \frac{1}{\varepsilon} [\mathcal{E}_I^\varepsilon + \mathcal{E}_I^{\varepsilon,\delta}]. \tag{5.15}$$

Integrating in time, we obtain in particular

$$\int_0^T \mathcal{E}_I^\varepsilon(t) \, dt \leq \varepsilon. \tag{5.16}$$

So suppose we have a smooth local solution to (5.3) with compact support of the flock Supp ρ_0. In Theorem 7.1 below, we demonstrate that such solutions can be obtained, for example, in the class

$$(\mathbf{u}, \rho) \in C_w([0, T); H^m \times (H^k \cap L^1_+)) \cap \text{Lip}([0, T); H^{m-1} \times (H^{k-1} \cap L^1_+)),$$

for $m \geq k + 1 > \frac{n}{2} + 2$. We consider what we call a *kinetic relative entropy* :

$$\eta^\varepsilon(t) = \frac{1}{2} \int_{\mathbb{R}^{2n}} |v - \mathbf{u}(x, t)|^2 f^\varepsilon(x, v, t) \, dx \, dv.$$

The role of this quantity is to control deviation of the distribution f^ε from limiting monokinetic ansatz. Our next goal is to establish smallness of η^ε on the entire time interval:

$$\eta^\varepsilon(t) \lesssim \varepsilon, \qquad \forall \delta \leq \varepsilon^2. \tag{5.17}$$

Before we write the equation for $\eta^\varepsilon(t)$, let us note that since all our solutions are smooth, we can treat all differential operations as classical. The macroscopic system for ε-solutions reads

$$
\begin{cases}
\rho_t^\varepsilon + \nabla \cdot (\rho^\varepsilon \mathbf{u}^\varepsilon) = 0, \\[2mm]
(\rho^\varepsilon \mathbf{u}^\varepsilon)_t + \nabla_x \cdot (\rho^\varepsilon \mathbf{u}^\varepsilon \otimes \mathbf{u}^\varepsilon + \mathcal{R}^\varepsilon) \\[2mm]
\qquad\qquad = \displaystyle\int_{\mathbb{R}^n} \rho^\varepsilon(x)\rho^\varepsilon(y)(\mathbf{u}^\varepsilon(y) - \mathbf{u}^\varepsilon(x))\phi(x-y)\,\mathrm{d}y \\[3mm]
\qquad\qquad\quad + \dfrac{1}{\varepsilon}\rho^\varepsilon(\mathbf{u}^{\varepsilon,\delta} - \mathbf{u}^\varepsilon).
\end{cases}
$$

Again, we recall that even if each individual component in the inertia term may not be smooth, the sum is

$$
\rho^\varepsilon \mathbf{u}^\varepsilon \otimes \mathbf{u}^\varepsilon + \mathcal{R}^\varepsilon = \int_{\mathbb{R}^n} (v \otimes v) f^\varepsilon(x, v, t)\,\mathrm{d}v.
$$

So let us break up $\eta^\varepsilon(t)$ into three components:

$$
\eta^\varepsilon(t) = \mathcal{E}^\varepsilon - \int_{\mathbb{R}^n} \rho^\varepsilon \mathbf{u}^\varepsilon \cdot \mathbf{u}\,\mathrm{d}x + \frac{1}{2}\int_{\mathbb{R}^n} \rho^\varepsilon |\mathbf{u}|^2\,\mathrm{d}x.
$$

As we already worked out the energy flux for $\mathcal{E}^\varepsilon$ above, let us move to the next two terms:

$$
\frac{\mathrm{d}}{\mathrm{d}t}\int_{\mathbb{R}^n} \rho^\varepsilon \mathbf{u}^\varepsilon \cdot \mathbf{u}\,\mathrm{d}x = \int_{\mathbb{R}^n} \partial_t(\rho^\varepsilon \mathbf{u}^\varepsilon) \cdot \mathbf{u}\,\mathrm{d}x + \int_{\mathbb{R}^n} \rho^\varepsilon \mathbf{u}^\varepsilon \cdot \partial_t \mathbf{u}\,\mathrm{d}x
$$

$$
= \int_{\mathbb{R}^n} (\rho^\varepsilon \mathbf{u}^\varepsilon \otimes \mathbf{u}^\varepsilon + \mathcal{R}^\varepsilon) : \nabla \mathbf{u}\,\mathrm{d}x
$$

$$
+ \frac{1}{2}\int_{\mathbb{R}^{2n}} \phi(x-y)\rho^\varepsilon(x)\rho^\varepsilon(y)(\mathbf{u}^\varepsilon(y) - \mathbf{u}^\varepsilon(x))(\mathbf{u}(x) - \mathbf{u}(y))\,\mathrm{d}y\,\mathrm{d}x
$$

$$
+ \frac{1}{\varepsilon}\int_{\mathbb{R}^n} \rho^\varepsilon(\mathbf{u}^{\varepsilon,\delta} - \mathbf{u}^\varepsilon) \cdot \mathbf{u}\,\mathrm{d}x - \int_{\mathbb{R}^n} \rho^\varepsilon \mathbf{u}^\varepsilon \otimes \mathbf{u} : \nabla \mathbf{u}\,\mathrm{d}x
$$

$$
+ \int_{\mathbb{R}^{2n}} \phi(x-y)\rho^\varepsilon(x)\rho(y)\mathbf{u}^\varepsilon(x)(\mathbf{u}(y) - \mathbf{u}(x))\,\mathrm{d}y\,\mathrm{d}x.
$$

$$
\frac{\mathrm{d}}{\mathrm{d}t}\frac{1}{2}\int_{\mathbb{R}^n} \rho^\varepsilon |\mathbf{u}|^2\,\mathrm{d}x = \int_{\mathbb{R}^n} \rho^\varepsilon \mathbf{u} \cdot \partial_t \mathbf{u}\,\mathrm{d}x + \frac{1}{2}\int_{\mathbb{R}^n} \partial_t \rho^\varepsilon |\mathbf{u}|^2\,\mathrm{d}x
$$

$$
= -\int_{\mathbb{R}^n} \rho^\varepsilon \mathbf{u} \otimes \mathbf{u} : \nabla \mathbf{u}\,\mathrm{d}x + \int_{\mathbb{R}^{2n}} \phi(x-y)\rho^\varepsilon(x)\rho(y)\mathbf{u}(x)(\mathbf{u}(y) - \mathbf{u}(x))\,\mathrm{d}y\,\mathrm{d}x
$$

$$
+ \int_{\mathbb{R}^n} \rho^\varepsilon \mathbf{u} \otimes \mathbf{u}^\varepsilon : \nabla \mathbf{u}\,\mathrm{d}x.
$$

All the inertia terms add up to

$$-\int_{\mathbb{R}^n} \rho^\varepsilon (\mathbf{u}^\varepsilon - \mathbf{u}) \otimes (\mathbf{u}^\varepsilon - \mathbf{u}) : \nabla \mathbf{u}\, dx \leq \|\nabla \mathbf{u}\|_\infty \int_{\mathbb{R}^n} \rho^\varepsilon |\mathbf{u}^\varepsilon - \mathbf{u}|^2\, dx.$$

What we see on the right-hand side is the macroscopic relative entropy which can be estimated by internal energy and kinetic relative entropy:

$$\int_{\mathbb{R}^n} \rho^\varepsilon |\mathbf{u}^\varepsilon - \mathbf{u}|^2\, dx = \int_{\mathbb{R}^n} |\mathbf{u}^\varepsilon - v + v - \mathbf{u}|^2 f^\varepsilon(x, v, t)\, dx\, dv \tag{5.18}$$
$$\lesssim \mathcal{E}_I^\varepsilon + \eta^\varepsilon.$$

Next, the Reynolds stress term is estimated by

$$\int_{\mathbb{R}^n} \mathcal{R}^\varepsilon : \nabla \mathbf{u}\, dx \leq \|\nabla \mathbf{u}\|_\infty \int_{\mathbb{R}^{2n}} |v - \mathbf{u}^\varepsilon(x, t)|^2 f^\varepsilon(x, v, t)\, dx\, dv \sim \mathcal{E}_I^\varepsilon.$$

As to the local alignment term, we use Lemma 5.1 and symmetry (5.8):

$$\int_{\mathbb{R}^n} \rho^\varepsilon (\mathbf{u}^{\varepsilon,\delta} - \mathbf{u}^\varepsilon) \cdot \mathbf{u}\, dx = \langle \mathbf{u}^{\varepsilon,\delta}, \mathbf{u} \rangle_{\rho^\varepsilon} - \langle \mathbf{u}^\varepsilon, \mathbf{u} \rangle_{\rho^\varepsilon} = \langle \mathbf{u}^\varepsilon, \mathbf{u}_{\rho^\varepsilon,\delta} \rangle_{\rho^\varepsilon} - \langle \mathbf{u}^\varepsilon, \mathbf{u} \rangle_{\rho^\varepsilon}$$
$$= \langle \mathbf{u}^\varepsilon, \mathbf{u}_{\rho^\varepsilon,\delta} - \mathbf{u} \rangle_{\rho^\varepsilon} \leq C \|\mathbf{u}^\varepsilon\|_\infty \delta \|\nabla \mathbf{u}\|_\infty \lesssim \delta.$$

So

$$\frac{1}{\varepsilon} \int_{\mathbb{R}^n} \rho^\varepsilon (\mathbf{u}^{\varepsilon,\delta} - \mathbf{u}^\varepsilon) \cdot \mathbf{u}\, dx \lesssim \frac{\delta}{\varepsilon}.$$

It remains to make estimates on the global alignment terms which is most involved. What helps to control these terms is in part the dissipation term $\mathcal{I}_2^\varepsilon$ coming from the energy inequality (5.15), the relative entropy itself η^ε, and the KR-distance between densities $W_1(\rho^\varepsilon, \rho)$; see (4.26).

First, we make a simple observation that $\mathcal{I}_2^\varepsilon$ dominates the corresponding macroscopic enstrophy:

$$\mathcal{I}_2^\varepsilon \geq \frac{1}{2} \int_{\mathbb{R}^{2n}} \rho^\varepsilon(x) \rho^\varepsilon(y) |\mathbf{u}^\varepsilon(x) - \mathbf{u}^\varepsilon(y)|^2 \phi(x - y)\, dy\, dx. \tag{5.19}$$

Indeed, expanding the $|w - v|^2$ term in $\mathcal{I}_2^\varepsilon$ we obtain

$$\mathcal{I}_2^\varepsilon = \int_{\mathbb{R}^{3n}} |v|^2 f^\varepsilon(x, v) \rho^\varepsilon(y) \phi(x - y)\, dy\, dx\, dv - \int_{\mathbb{R}^{2n}} \rho^\varepsilon(x) \rho^\varepsilon(y) \mathbf{u}^\varepsilon(x) \mathbf{u}^\varepsilon(y)\, dx\, dy.$$

Expanding the macroscopic enstrophy, we obtain

$$\frac{1}{2}\int_{\mathbb{R}^{2n}}\rho^{\varepsilon}(x)\rho^{\varepsilon}(y)|\mathbf{u}^{\varepsilon}(x)-\mathbf{u}^{\varepsilon}(y)|^{2}\phi(x-y)\,dy\,dx$$

$$=\int_{\mathbb{R}^{2n}}\rho^{\varepsilon}(x)\rho^{\varepsilon}(y)|\mathbf{u}^{\varepsilon}(x)|^{2}\phi(x-y)\,dy\,dx$$

$$-\int_{\mathbb{R}^{2n}}\rho^{\varepsilon}(x)\rho^{\varepsilon}(y)\mathbf{u}^{\varepsilon}(x)\mathbf{u}^{\varepsilon}(y)\,dx\,dy.$$

However, the total energy density dominates the macroscopic one:

$$\int_{\mathbb{R}^{n}}|v|^{2}f^{\varepsilon}(x,v)\,dv=\int_{\mathbb{R}^{n}}|v-\mathbf{u}^{\varepsilon}(x)|^{2}f^{\varepsilon}(x,v)\,dv+\rho^{\varepsilon}(x)|\mathbf{u}^{\varepsilon}(x)|^{2}\geq\rho^{\varepsilon}(x)|\mathbf{u}^{\varepsilon}(x)|^{2}.$$

This proves (5.19).

First, let us collect all the alignment terms:

$$A=\int_{\mathbb{R}^{2n}}\phi(x-y)\rho^{\varepsilon}(x)\rho(y)(\mathbf{u}(x)-\mathbf{u}^{\varepsilon}(x))(\mathbf{u}(y)-\mathbf{u}(x))\,dy\,dx$$

$$+\frac{1}{2}\int_{\mathbb{R}^{2n}}\phi(x-y)\rho^{\varepsilon}(x)\rho^{\varepsilon}(y)(\mathbf{u}^{\varepsilon}(x)-\mathbf{u}^{\varepsilon}(y))(\mathbf{u}(x)-\mathbf{u}(y))\,dy\,dx.$$

For the second term, we use dissipation (5.19) to partially absorb it:

$$\frac{1}{2}\int_{\mathbb{R}^{2n}}\phi(x-y)\rho^{\varepsilon}(x)\rho^{\varepsilon}(y)(\mathbf{u}^{\varepsilon}(x)-\mathbf{u}^{\varepsilon}(y))(\mathbf{u}(x)-\mathbf{u}(y))\,dy\,dx$$

$$=\frac{1}{2}\int_{\mathbb{R}^{2n}}\phi(x-y)\rho^{\varepsilon}(x)\rho^{\varepsilon}(y)|\mathbf{u}^{\varepsilon}(x)-\mathbf{u}^{\varepsilon}(y)|^{2}\,dy\,dx$$

$$+\int_{\mathbb{R}^{2n}}\phi(x-y)\rho^{\varepsilon}(x)\rho^{\varepsilon}(y)(\mathbf{u}^{\varepsilon}(x)-\mathbf{u}^{\varepsilon}(y))(\mathbf{u}(x)-\mathbf{u}^{\varepsilon}(x))\,dy\,dx.$$

So the total sum of alignments is bounded by

$$A\leq I_{2}^{\varepsilon}+\int_{\mathbb{R}^{2n}}\phi(x-y)\rho^{\varepsilon}(x)\rho(y)(\mathbf{u}(x)-\mathbf{u}^{\varepsilon}(x))(\mathbf{u}(y)-\mathbf{u}(x))\,dy\,dx$$

$$+\int_{\mathbb{R}^{2n}}\phi(x-y)\rho^{\varepsilon}(x)\rho^{\varepsilon}(y)(\mathbf{u}^{\varepsilon}(x)-\mathbf{u}^{\varepsilon}(y))(\mathbf{u}(x)-\mathbf{u}^{\varepsilon}(x))\,dy\,dx$$

$$=I_{2}^{\varepsilon}+\int_{\mathbb{R}^{2n}}\phi(x-y)\rho^{\varepsilon}(x)(\rho(y)-\rho^{\varepsilon}(y))(\mathbf{u}(x)-\mathbf{u}^{\varepsilon}(x))(\mathbf{u}(y)-\mathbf{u}(x))\,dy\,dx$$

$$+\int_{\mathbb{R}^{2n}}\phi(x-y)\rho^{\varepsilon}(x)\rho^{\varepsilon}(y)(\mathbf{u}^{\varepsilon}(x)-\mathbf{u}(x)+\mathbf{u}(y)-\mathbf{u}^{\varepsilon}(y))(\mathbf{u}(x)-\mathbf{u}^{\varepsilon}(x))\,dy\,dx.$$

The last term is bounded by the macroscopic relative entropy which from (5.18) is bounded by $\mathcal{E}_{I}^{\varepsilon}+\eta^{\varepsilon}$. Let us collect the obtained estimates so far:

$$\frac{d}{dt}\eta^{\varepsilon} \lesssim \mathcal{E}_I^{\varepsilon} + \eta^{\varepsilon} + \frac{\delta}{\varepsilon}$$

$$+ \int_{\mathbb{R}^{2n}} \phi(x-y)\rho^{\varepsilon}(x)(\rho(y) - \rho^{\varepsilon}(y))(\mathbf{u}(x) - \mathbf{u}^{\varepsilon}(x))(\mathbf{u}(y) - \mathbf{u}(x))\, dy\, dx.$$

It remains to estimate the last alignment term. We break it up as follows:

$$\int_{\mathbb{R}^n} \rho^{\varepsilon}(x)(\mathbf{u}(x) - \mathbf{u}^{\varepsilon}(x)) \left[\int_{\mathbb{R}^n} \mathbf{u}(y)(\rho(y) - \rho^{\varepsilon}(y))\phi(x-y)\, dy \right] dx$$

$$- \int_{\mathbb{R}^n} \rho^{\varepsilon}(x)(\mathbf{u}(x) - \mathbf{u}^{\varepsilon}(x))\mathbf{u}(x) \left[\int_{\mathbb{R}^n} (\rho(y) - \rho^{\varepsilon}(y))\phi(x-y)\, dy \right] dx.$$

The inner integrals are clearly bounded by constant multiples of $W_1(\rho^{\varepsilon}, \rho)$ and the remaining outer integral by $\sqrt{\eta^{\varepsilon}}$. So we obtain

$$\frac{d}{dt}\eta^{\varepsilon} \lesssim \mathcal{E}_I^{\varepsilon} + \eta^{\varepsilon} + \frac{\delta}{\varepsilon} + W_1(\rho^{\varepsilon}, \rho)\sqrt{\eta^{\varepsilon}}. \tag{5.20}$$

Lemma 5.2 *We have the following bound for all $t < T$:*

$$W_1(\rho^{\varepsilon}(t), \rho(t)) \lesssim \varepsilon + \left[\int_0^t \eta^{\varepsilon}(s)\, ds \right]^{1/2}.$$

Assuming the lemma holds, we obtain

$$\frac{d}{dt}\eta^{\varepsilon} \lesssim \mathcal{E}_I^{\varepsilon} + \eta^{\varepsilon} + \frac{\delta}{\varepsilon} + \varepsilon\sqrt{\eta^{\varepsilon}} + \sqrt{\eta^{\varepsilon}}\left[\int_0^t \eta^{\varepsilon}(s)\, ds \right]^{1/2}$$

$$\leq \mathcal{E}_I^{\varepsilon} + \eta^{\varepsilon} + \frac{\delta}{\varepsilon} + \varepsilon^2 + \eta^{\varepsilon} + \int_0^t \eta^{\varepsilon}(s)\, ds.$$

Integrating to t and using (5.16), we obtain

$$\eta^{\varepsilon}(t) \lesssim \eta^{\varepsilon}(0) + \varepsilon + \frac{\delta}{\varepsilon} + \int_0^t \eta^{\varepsilon}(s)\, ds.$$

At this point, we see that $\delta \sim \varepsilon^2$ is the optimal value of parameter δ which gives a rate of ε. Given that $d(f_0^{\varepsilon}, f_0) \leq \varepsilon$, we have

$$\eta^{\varepsilon}(0) = \frac{1}{2}\int_{\mathbb{R}^{2n}} |v - \mathbf{u}_0(x)|^2[f_0^{\varepsilon}(x, v)\, dx\, dv - d f_0(x, v)] \leq C W_1(f_0^{\varepsilon}, f_0) \leq \varepsilon.$$

So

$$\eta^{\varepsilon}(t) \lesssim \varepsilon + \int_0^t \eta^{\varepsilon}(s)\, ds.$$

Grönwall's lemma establishes that

$$\eta^\varepsilon(t) \lesssim \varepsilon.$$

Again, in view of Lemma 5.2, this also shows that

$$W_1(\rho^\varepsilon(t), \rho(t)) \lesssim \sqrt{\varepsilon}.$$

To establish convergence to monokinetic ansatz, we fix $g \in \mathrm{Lip}(\mathbb{R}^{2n})$, $\|\nabla g\|_\infty \le 1$, and compute

$$\int_{\mathbb{R}^{2n}} g(x, v)[f^\varepsilon(x, v)\, dx\, dv - df(x, v)] = \int_{\mathbb{R}^{2n}} g(x, v) f^\varepsilon(x, v)\, dx\, dv$$

$$- \int_{\mathbb{R}^n} g(x, \mathbf{u}(x))\rho(x)\, dx$$

$$= \int_{\mathbb{R}^{2n}} (g(x, v) - g(x, \mathbf{u}(x))) f^\varepsilon(x, v)\, dx\, dv + \int_{\mathbb{R}^n} g(x, \mathbf{u}(x))(\rho^\varepsilon(x) - \rho(x))\, dx$$

$$\le \int_{\mathbb{R}^{2n}} |v - \mathbf{u}(x)| f^\varepsilon(x, v)\, dx\, dv + C W_1(\rho^\varepsilon(t), \rho(t)) \lesssim \sqrt{\eta^\varepsilon} + \sqrt{\varepsilon} \lesssim \sqrt{\varepsilon}.$$

This proves the result.

Proof (Proof of Lemma 5.2) In the course of the proof, we will use the two characteristic flow maps (X, V) corresponding to monokinetic solution and $(X^\varepsilon, V^\varepsilon)$ corresponding to f^ε. Note that the monokinetic flow is smooth as it corresponds to a measure-valued solution of the classical kinetic Cucker-Smale equation; see Sect. 4.3. The flow $(X^\varepsilon, V^\varepsilon)$ is also smooth; however, it is unstable as $\varepsilon \to 0$. So it is prohibited, nor do we require, to use any regularity estimates on it.

Let us consider another distribution, which is the push-forward of f_0^ε under the monokinetic flow:

$$\tilde{f}^\varepsilon = (X, V)\#f_0^\varepsilon.$$

Let us fix $g \in \mathrm{Lip}(\mathbb{R}^n)$, $\|\nabla g\|_\infty \le 1$, and compute

$$\int_{\mathbb{R}^n} g(x)[\rho^\varepsilon(x) - \rho(x)]\, dx = \int_{\mathbb{R}^{2n}} g(x)[f^\varepsilon(x, v)\, dx\, dv - df(x, v)]$$

$$= \int_{\mathbb{R}^{2n}} g(x)[f^\varepsilon(x, v) - \tilde{f}^\varepsilon(x, v)]\, dx\, dv + \int_{\mathbb{R}^{2n}} g(x)[\tilde{f}^\varepsilon(x, v)\, dx\, dv - df(x, v)]$$

$$= \int_{\mathbb{R}^{2n}} [g(X^\varepsilon(\omega, t)) - g(X(\omega, t))] f_0^\varepsilon(\omega)\, d\omega + \int_{\mathbb{R}^{2n}} g(X(\omega, t))[f_0^\varepsilon(\omega)\, d\omega - df_0]$$

$$\le \|\nabla g\|_\infty \int_{\mathbb{R}^{2n}} |X^\varepsilon(\omega, t) - X(\omega, t)| f_0^\varepsilon(\omega)\, d\omega + C\varepsilon.$$

Denote

$$D = \int_{\mathbb{R}^{2n}} |X^\varepsilon(\omega, t) - X(\omega, t)| f_0^\varepsilon(\omega)\, d\omega.$$

Differentiating, we obtain

$$\frac{d}{dt} D \leq \int_{\mathbb{R}^{2n}} |V^\varepsilon(\omega, t) - V(\omega, t)| f_0^\varepsilon(\omega)\, d\omega$$

$$\leq \int_{\mathbb{R}^{2n}} |V^\varepsilon(\omega, t) - \mathbf{u}(X^\varepsilon(\omega, t), t)| f_0^\varepsilon(\omega)\, d\omega$$

$$+ \int_{\mathbb{R}^{2n}} |\mathbf{u}(X^\varepsilon(\omega, t), t) - \mathbf{u}(X(\omega, t), t)| f_0^\varepsilon(\omega)\, d\omega$$

$$+ \int_{\mathbb{R}^{2n}} |\mathbf{u}(X(\omega, t), t) - V(\omega, t)| f_0^\varepsilon(\omega)\, d\omega$$

$$\leq \int_{\mathbb{R}^{2n}} |v - \mathbf{u}(x, t)| f^\varepsilon(x, v, t)\, dx\, dv + \|\nabla \mathbf{u}\|_\infty D$$

$$+ \int_{\mathbb{R}^{2n}} |\mathbf{u}(X(\omega, t), t) - V(\omega, t)| [f_0^\varepsilon(\omega)\, d\omega - d f_0(\omega)]$$

$$+ \int_{\mathbb{R}^{2n}} |\mathbf{u}(X(\omega, t), t) - V(\omega, t)|\, d f_0(\omega)$$

$$\lesssim \sqrt{\eta^\varepsilon} + D + W_1(f_0^\varepsilon, f_0) + 0,$$

the latter due to the monokinetic assumption:

$$\int_{\mathbb{R}^{2n}} |\mathbf{u}(X(\omega, t), t) - V(\omega, t)|\, d f_0(\omega) = \int_{\mathbb{R}^{2n}} |\mathbf{u}(x, t) - v|\, d f(\omega, t) = 0.$$

We thus obtain

$$\frac{d}{dt} D \leq \sqrt{\eta^\varepsilon} + D + \varepsilon.$$

Since $D(0) = 0$, the lemma follows by Grönwall's inequality. $\qquad\square$

5.3 Notes and References

Hydrodynamic limit in the context of alignment dynamics was first addressed in Kang and Vasseur [57] where a partial result was established without the CS alignment term. Later Figalli and Kang [41] extended the result to the full model considering local alignment with rough macroscopic velocity:

$$\partial_t f^\varepsilon + v \cdot \nabla_x f^\varepsilon + \lambda \nabla_v \cdot [f^\varepsilon F(f^\varepsilon)] + \frac{1}{\varepsilon} \nabla_v [f^\varepsilon (\mathbf{u}^\varepsilon - v)] = 0. \tag{5.21}$$

Both works are carried out in the framework of weak solutions to (5.21) previously constructed by Karper, Mellet, and Trivisa [58] for a much wider class of models. The approach is based upon analysis of the macroscopic relative entropy which necessitates a study of characteristic flow to the rough field $\mathbf{u}^\varepsilon$. The theorem is proved on the torus for no-vacuum solutions as a result of limitations that come from dealing with macroscopic values. Following the same methodology but with the use of kinetic relative entropy allows to bypass those limitations, so the result presented here is more general. The use of the Favre filtration also bridges the gap between kinetic and discrete systems as we can justify such a kinetic formulation via a mean-field limit.

A parallel approach was developed in the trilogy of papers by Karper, Mellet, and Trivisa [58–60]. One starts from the stochastically forced hybrid CS-MT model:

$$\begin{cases} \dot{\mathbf{x}}_i = \mathbf{v}_i, \\[2mm] \dot{\mathbf{v}}_i = \displaystyle\sum_{j=1}^{N} m_j \phi(\mathbf{x}_i - \mathbf{x}_j)(\mathbf{v}_j - \mathbf{v}_i) \\[4mm] \qquad + \dfrac{\sigma}{\sum_{j=1}^{N} m_j \psi_\delta(\mathbf{x}_i - \mathbf{x}_j)} \displaystyle\sum_{j=1}^{N} m_j \psi_\delta(\mathbf{x}_i - \mathbf{x}_j)(\mathbf{v}_j - \mathbf{v}_i) \\[4mm] \qquad + \sqrt{2\sigma}\, \dot{W}_i, \end{cases}$$

which results in the kinetic Vlasov-Fokker-Planck model:

$$f_t + v \cdot \nabla_x f + \nabla_v(f F(f)) + \sigma \nabla_v(f(\mathbf{u} - v)) = \sigma \Delta_v f. \tag{5.22}$$

In the strong diffusion and local alignment limit $\sigma \to \infty$, solutions relax to the Maxwellian

$$f(x, v, t) = \rho(x, t) e^{-|\mathbf{u}(x,t)-v|^2}. \tag{5.23}$$

Reading off the system for macroscopic quantities, one arrives at an isentropic EAS:

$$\begin{cases} \rho_t + \nabla \cdot (\rho \mathbf{u}) = 0, \\[3mm] (\rho \mathbf{u})_t + \nabla_x \cdot (\rho \mathbf{u} \otimes \mathbf{u}) + \nabla p = \displaystyle\int_{\mathbb{R}^n} \rho(x)\rho(y)(\mathbf{u}(y) - \mathbf{u}(x))\phi(x - y)\,dy \end{cases}$$

with isothermal pressure law $p = \rho$. This system received much less attention than its pressureless counterpart; see the partial results for smooth [25, 60] and singular [29] models. A distinct feature of pressured dynamics is that solutions, say, on the torus, converge to purely uniform distributions $\rho = M$, as those present the only possible outcome for a fully aligned flock. For singular models, we already commented the very issue of defining $F(f)$ becomes central. For weakly singular kernels $\beta < n$, a valuable insight into hydrodynamic limit was offered by Poyato and Soler [86], who looked into the model with friction, diffusion, and external forcing:

$$\varepsilon f_t + \varepsilon v \cdot \nabla_x f + \nabla_x \psi \cdot \nabla_v f + \nabla_v (f F_\varepsilon(f)) = \Delta_v f + \nabla_v (f v),$$

where ψ is an external potential and F_ε is defined as before with desingularized kernel:

$$\phi_\varepsilon(r) = \frac{1}{(\varepsilon^2 + r^2)^{\beta/2}}.$$

The hydrodynamic limit as $\varepsilon \to 0$ removes inertia terms, and one obtains the following system:

$$\rho_t + \nabla \cdot (\rho \mathbf{u}) = 0, \quad \nabla \psi + \mathbf{u} = \phi * (\rho \mathbf{u}) - \mathbf{u} \phi * \rho.$$

Note that the dynamics here is driven by the external potential, as if $\psi = $ const, then the system has only trivial traveling wave solutions $\mathbf{u} \equiv \bar{\mathbf{u}}$, $\rho = \rho(x - t\bar{\mathbf{u}})$.

Chapter 6
Euler Alignment System

In this chapter we will focus on the analysis of long time behavior of the pressureless Euler alignment system:

$$
\begin{cases}
\rho_t + \nabla \cdot (\rho \mathbf{u}) = 0, \\[2mm]
\mathbf{u}_t + \mathbf{u} \cdot \nabla \mathbf{u} = \displaystyle\int_{\Omega} \phi(x, y)(\mathbf{u}(y) - \mathbf{u}(x))\rho(y)\,\mathrm{d}y,
\end{cases}
\qquad (x, t) \in \Omega \times \mathbb{R}_+,
$$

(6.1)

where $\Omega = \mathbb{T}^n$ or $\mathbb{R}^n$. We will recast the basic flocking results, similar to the micro- and meso-systems, in the context of the hydrodynamic system (6.1) with convolution-type communication. Our further focus will shift to local kernels and demonstrating capabilities and limitations of the spectral method. This will motivate the introduction of the topological models later in Sect. 6.4.

We start with basic energetics of the system and straightforward extensions of the results obtained in the discrete case.

6.1 Basic Properties. Energy Law

For symmetric kernels, just as in the discrete case, the system (6.1) conserves mass and momentum:

$$
\overline{\rho \mathbf{u}} = \int_{\Omega} \rho \mathbf{u}\,\mathrm{d}x, \quad M = \int_{\Omega} \rho(x, t)\,\mathrm{d}x.
$$

This allows to predict the limiting alignment velocity to be $\bar{\mathbf{u}} = \frac{1}{M}\overline{\rho \mathbf{u}}$. For convolution kernels, $\phi(x, y) = \phi(x - y)$, the system is also Galilean invariant. If this is the case, we can always assume that $\bar{\mathbf{u}} = \mathbf{0}$.

© Springer Nature Switzerland AG 2021
R. Shvydkoy, *Dynamics and Analysis of Alignment Models of Collective Behavior*,
Nečas Center Series, https://doi.org/10.1007/978-3-030-68147-0_6

The crucial feature of the alignment term in (6.1) is its commutator representation given by

$$C_\phi(\mathbf{u}, \rho) = \mathcal{L}_\phi(\rho \mathbf{u}) - \mathcal{L}_\phi(\rho)\mathbf{u}, \tag{6.2}$$

where $\mathcal{L}_\phi$ can take two different forms: either

$$\mathcal{L}_\phi(f)(x) = \int_\Omega \phi(x, y) f(y) \, dy, \tag{6.3}$$

suitable for smooth kernels or

$$\mathcal{L}_\phi(f)(x) = \int_\Omega \phi(x, y)(f(y) - f(x)) \, dy, \tag{6.4}$$

suitable for singular kernels. Note that for smooth kernels of convolution type, we have $\mathcal{L}_\phi f = \phi * f$.

The kinetic energy is given by

$$\mathcal{E} = \frac{1}{2} \int_\Omega \rho |\mathbf{u}|^2 \, dx.$$

If $\bar{\mathbf{u}} = \mathbf{0}$, the energy becomes a measure of alignment due to its relation to the L^2-fluctuation functional given by

$$\mathcal{E} = \frac{1}{2M} \mathcal{V}_2, \quad \mathcal{V}_2(t) = \frac{1}{2} \int_{\Omega \times \Omega} |\mathbf{u}(x, t) - \mathbf{u}(y, t)|^2 \rho(x, t) \rho(y, t) \, dx \, dy.$$

As in the discrete case, we mostly work with the fluctuation functionals to make the argument independent of the Galilean invariance of the model. Thus, using just symmetry of the kernel, we obtain the basic energy law:

$$\frac{d}{dt} \mathcal{V}_2 = -2M\mathcal{I}_2,$$

$$\mathcal{I}_2 = \frac{1}{2} \int_{\Omega \times \Omega} \phi(x, y) |\mathbf{u}(y) - \mathbf{u}(x)|^2 \rho(y) \rho(x) \, dy. \tag{6.5}$$

Since all the macroscopic quantities are measured relative to mass, it is natural to introduce the density measure:

$$dm_t = \rho(x, t) \, dx.$$

In view of the transport nature of the continuity equation, this measure is transported along the flow of $\mathbf{u}$. Namely, if

$$\partial_t X(\alpha, t, t_0) = \mathbf{u}(X(\alpha, t, t_0), t), \quad t > t_0,$$

$$X(\alpha, t_0, t_0) = \alpha,$$

then m_t is a push-forward of m_{t_0} under $X(\cdot, t, t_0)$:

$$m_t = X(\cdot, t, t_0) \# m_{t_0}. \tag{6.6}$$

In other words, for any g

$$\int_{\mathbb{R}^n} g(X(\alpha, t, t_0)) \, dm_{t_0}(\alpha) = \int_{\mathbb{R}^n} g(x) \, dm_t(x). \tag{6.7}$$

We also denote $X(\cdot, t, 0) = X(\cdot, t)$. In particular, the density support is transported by the flow:

$$\operatorname{Supp} \rho(t) = X(\operatorname{Supp} \rho_0, t).$$

Denoting the Lagrangian velocity by

$$\mathbf{v}(\alpha, t) = \mathbf{u}(X(\alpha, t), t)$$

for short and denoting

$$\mathbf{v}_{\alpha\beta} = \mathbf{v}(\alpha, t) - \mathbf{v}(\beta, t), \quad \phi_{\alpha\beta} = \phi(X(\alpha, t) - X(\beta, t)),$$

we can rewrite the velocity equation as

$$\partial_t \mathbf{v}(\alpha, t) = \int_\Omega \phi_{\alpha\beta} [\mathbf{v}(\beta, t) - \mathbf{v}(\alpha, t)] \, dm_0(\beta). \tag{6.8}$$

The L^p-fluctuations are defined by

$$\mathcal{V}_p(t) = \frac{1}{p} \int_{\Omega \times \Omega} |\mathbf{v}(\beta, t) - \mathbf{v}(\alpha, t)|^p \, dm_0(\alpha, \beta),$$

where $dm_0(\alpha, \beta) = dm_0(\alpha) \times dm_0(\beta)$.

We can see that these structures bare close resemblance with the discrete and kinetic settings, and as a result, in Lagrangian coordinates many computations become very similar too. For example, it is straightforward to see that all $\mathcal{V}_p$'s are decaying in time and (2.42) translates into just

$$\frac{d}{dt} \mathcal{V}_p \leq 0.$$

Another fundamental property of the system (6.1) is the maximum principle for each scalar velocity component $\ell(\mathbf{u})$, $\ell \in (\mathbb{R}^n)^*$, provided the maxima are achieved, which is of course true on a periodic domain or if $\bar{\mathbf{u}} = \mathbf{0}$ and the solution decays at infinity on $\mathbb{R}^n$.

6.2 Hydrodynamic Flocking and Stability

We assume here that the domain is $\mathbb{R}^n$ as the discussion carries over to periodic settings directly and ϕ is of convolution type unless noted otherwise.

Let us consider a flock of finite diameter and a compact domain Ω containing Supp ρ_0. We define the flock parameters as follows

$$\mathcal{D}_\Omega(t) = \max_{\alpha,\beta \in \Omega} |X(\alpha, t) - X(\beta, t)|, \quad \mathcal{A}_\Omega(t) = \max_{\alpha,\beta \in \Omega} |\mathbf{v}(\alpha, t) - \mathbf{v}(\beta, t)|.$$

Just as in the kinetic case, it is important to consider this more general setup in order to have a capability to compare flocks and study stability. Since the domain Ω is fixed for all time, one can apply Rademacher's Lemma 2.2 and carry out the same argument as in the discrete and kinetic cases. Thus, for general nonnegative kernels, we have

$$\frac{d}{dt} \mathcal{A}_\Omega(t) \le 0. \tag{6.9}$$

By expanding Ω to $\mathbb{R}^n$, we also see that the global amplitude is not increasing either:

$$\mathcal{A}(t) = \sup_{\alpha,\beta \in \mathbb{R}^n} |\mathbf{v}(\alpha, t) - \mathbf{v}(\beta, t)|, \quad \mathcal{A}(t) \le \mathcal{A}(0). \tag{6.10}$$

For heavy tail kernels, we have the following result whose proof is identical to Theorem 2.2.

Theorem 6.1 *The system* (6.1) *aligns and flocks exponentially fast provided ϕ is nonincreasing, everywhere positive, and satisfies the heavy tail condition* (2.6). *More precisely, if $\overline{\mathcal{D}}_\Omega$ solves* (2.20), *then we have the following estimates:*

$$\sup_{t \ge 0} \mathcal{D}_\Omega(t) = \overline{\mathcal{D}}_\Omega, \quad \mathcal{A}_\Omega(t) \le \mathcal{A}_\Omega(0) e^{-\lambda M \phi(\overline{\mathcal{D}}_\Omega) t}. \tag{6.11}$$

The corresponding version of Theorem 6.1 for the Motsch-Tadmor model (2.53) carries over from the discrete context (2.53) verbatim. For systems with degenerate communication, the corrector method yields the same statement as in Theorem 2.6(ii), which was proved independent of the number of agents. Here one makes use of macroscopic quantities throughout:

$$d_{\alpha\beta} = -X_{\alpha\beta} \cdot \frac{\mathbf{v}_{\alpha\beta}}{|\mathbf{v}_{\alpha\beta}|}, \quad \mathcal{G}_3 = \int_{\mathbb{R}^{2n}} |\mathbf{v}_{\alpha\beta}|^3 \psi(d_{\alpha\beta}) \chi(|X_{\alpha\beta}|) \, \mathrm{d}m_0(\alpha, \beta), \quad \text{etc.}$$

The Lagrangian formulation of the Euler Alignment system is given by

$$\partial_t X(\alpha, t) = \mathbf{v}(\alpha, t), \tag{6.12}$$

$$\partial_t \mathbf{v}(\alpha, t) = \lambda \int_{\mathbb{R}^n} \phi(X(\alpha, t) - X(\beta, t))[\mathbf{v}(\beta, t) - \mathbf{v}(\alpha, t)] \rho_0(\beta) \, \mathrm{d}\beta. \tag{6.13}$$

As we can see, it has almost identical structure to its kinetic counterpart (4.14). This allows us to obtain similar stability and regularity estimates for smooth-type communication kernels. We start with estimates on the deformation, and it will be useful to reproduce them pointwise. So, we have

$$\partial_t \nabla X(\alpha, t) = \nabla \mathbf{v}(\alpha, t),$$

$$\partial_t \nabla \mathbf{v}(\alpha, t) = \lambda \int_{\mathbb{R}^n} \nabla^\top X(\alpha, t) \nabla \phi(X(\alpha, t) - X(\beta, t))$$

$$\otimes (\mathbf{v}(\beta, t) - \mathbf{v}(\alpha, t)) \rho_0(\beta) \, \mathrm{d}\beta$$

$$- \lambda \nabla \mathbf{v}(\alpha, t) \int_{\mathbb{R}^n} \phi(X(\alpha, t) - X(\beta, t)) \rho_0(\beta) \, \mathrm{d}\beta.$$

In the case of local kernels, we do not hope for good long time control. So, we simply note that $\mathcal{A}_\Omega(t) \leq \mathcal{A}_\Omega(0)$ for any domain $\Omega \subset \mathbb{R}^n$ due to (6.9) and the kernel is bounded. This implies

$$\|\nabla \mathbf{v}(t)\|_{L^\infty(\mathbb{R}^n)} + \|\nabla X(t)\|_{L^\infty(\mathbb{R}^n)} \leq C_1 e^{C_2 t}. \tag{6.14}$$

For heavy tail kernels, we can deduce better long time estimates on every fixed compact domain Ω containing Supp ρ_0. So, let us fix such Ω and deduce as in the kinetic case:

$$\frac{\mathrm{d}}{\mathrm{d}t} \|\nabla X\|_{L^\infty(\Omega)} \leq \|\nabla \mathbf{v}\|_{L^\infty(\Omega)},$$

$$\frac{\mathrm{d}}{\mathrm{d}t} \|\nabla \mathbf{v}\|_{L^\infty(\Omega)} \leq \lambda \|\nabla \phi\|_\infty \|\nabla X\|_{L^\infty(\Omega)} \mathcal{A}_\Omega(0) e^{-t\lambda M \phi(\overline{\mathcal{D}}_\Omega)}$$

$$- \lambda M \phi(\overline{\mathcal{D}}_\Omega) \|\nabla \mathbf{v}\|_{L^\infty(\Omega)}.$$

The resulting estimate (2.31) implies, noting that initially $\nabla X = \mathbb{I}$, $\nabla \mathbf{v} = \nabla \mathbf{u}_0$

$$a\|\nabla X(t)\|^2_{L^\infty(\Omega)} + e^{bt} \|\nabla \mathbf{v}(t)\|^2_{L^\infty(\Omega)} \leq \frac{4\sqrt{a}}{b}(a + \|\nabla \mathbf{u}_0\|^2_{L^\infty(\Omega)}), \tag{6.15}$$

where $a = \lambda M \mathcal{A}_\Omega(0)$ and $b = \lambda M \phi(\overline{\mathcal{D}}_\Omega)$. In particular, we can see that $\|\nabla \mathbf{v}(t)\|^2_{L^\infty(\Omega)}$ decays exponentially fast.

Suppose now we have two flocks with initial densities ρ'_0, ρ''_0. It turns out that our kinetic result can be carried out almost verbatim in the hydrodynamic setting with the use of Kantorovich-Rubinstein distance on $\mathcal{M}_+(\mathbb{R}^n)$. So, if Ω is a compact domain enclosing both initial flocks, i.e.

$$\text{Supp } \rho'_0 \cup \text{Supp } \rho''_0 \subset \Omega,$$

and the masses and momenta are equal, $M' = M''$, $\bar{\mathbf{v}}' = \bar{\mathbf{v}}''$, then we obtain

$$\frac{d}{dt} \|X'(t) - X''(t)\|_{L^\infty(\Omega)} \le \|\mathbf{v}'(t) - \mathbf{v}''(t)\|_{L^\infty(\Omega)},$$

$$\frac{d}{dt} \|\mathbf{v}'(t) - \mathbf{v}''(t)\|_{L^\infty(\Omega)} \le \lambda \|\phi\|_{W^{1,\infty}} (\|\nabla X'\|_{L^\infty(\Omega)} \mathcal{A}'_\Omega(t) + \|\nabla \mathbf{v}'\|_{L^\infty(\Omega)})$$

$$\times W_1(\rho'_0, \rho''_0)$$

$$+ 2\lambda M \|\nabla \phi\|_\infty \|X'(t) - X''(t)\|_{L^\infty(\Omega)} \mathcal{A}'_\Omega(t)$$

$$- \lambda M \phi(\mathcal{D}''_\Omega) \|\mathbf{v}'(t) - \mathbf{v}''(t)\|_{L^\infty(\Omega)}.$$

Subsequently, with the use of (6.14) and (6.15), and recalling that characteristics start from the same values and $\|\mathbf{v}'(0) - \mathbf{v}''(0)\|_{L^\infty(\Omega)} = \|\mathbf{u}'_0 - \mathbf{u}''_0\|_{L^\infty(\Omega)}$, we obtain for general kernels

$$\|X'(t) - X''(t)\|_{L^\infty(\Omega)} + \|\mathbf{v}'(t) - \mathbf{v}''(t)\|_{L^\infty(\Omega)}$$

$$\le C e^{Ct} [W_1(\rho'_0, \rho''_0) + \|\mathbf{u}'_0 - \mathbf{u}''_0\|_{L^\infty(\Omega)}],$$

while for kernels with heavy tail

$$\|X'(t) - X''(t)\|_{L^\infty(\Omega)} \le C[W_1(\rho'_0, \rho''_0) + \|\mathbf{u}'_0 - \mathbf{u}''_0\|_{L^\infty(\Omega)}], \tag{6.16}$$

$$\|\mathbf{v}'(t) - \mathbf{v}''(t)\|_{L^\infty(\Omega)} \le C e^{-ct} [W_1(\rho'_0, \rho''_0) + \|\mathbf{u}'_0 - \mathbf{u}''_0\|_{L^\infty(\Omega)}]. \tag{6.17}$$

Continuing with the same KR-distance computation leading to (4.29)–(4.30), we obtain for general kernels

$$W_1(\rho'(t), \rho''(t)) \le C e^{Ct} [W_1(\rho'_0, \rho''_0) + \|\mathbf{u}'_0 - \mathbf{u}''_0\|_{L^\infty(\Omega)}], \tag{6.18}$$

and for heavy tail kernels

$$W_1(\rho'(t), \rho''(t)) \le C[W_1(\rho'_0, \rho''_0) + \|\mathbf{u}'_0 - \mathbf{u}''_0\|_{L^\infty(\Omega)}]. \tag{6.19}$$

Remark 6.1 On the real line $\mathbb{R}$, the KR-distance between two densities ρ', ρ'' is equal to the L^1-norm of the difference of the corresponding cumulative distribution

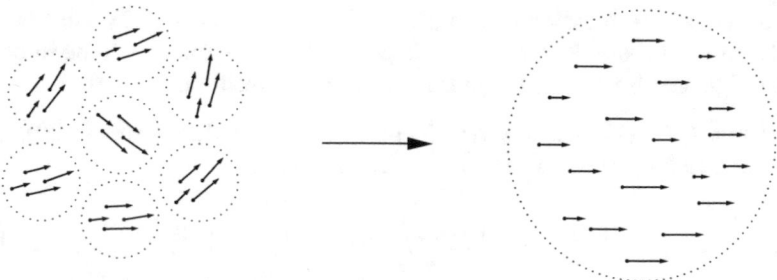

Fig. 6.1 Emerging global behavior from local alignment

functions $F'(x) = \int_{-\infty}^{x} \rho'(y)\,dy$, $F''(x) = \int_{-\infty}^{x} \rho''(y)\,dy$:

$$W_1(\rho', \rho'') = \|F' - F''\|_{L^1}.$$

See [103] for details.

6.3 Spectral Method. Hydrodynamic Connectivity

In the next two section, we will address the fundamental question: Can alignment still occur in systems with purely local interactions? Can we see transient clustering before global alignment emerges; see Fig. 6.1?

It is clear that the role of connectivity in addressing this question becomes more prominent. We already saw in Sect. 2.3 that in the open space, two disconnected flocks can disperse in different directions without ever aligning. In a confined environment, such as periodic domain, connectivity can be expressed by a lower bound on the density as a function of time $\rho > c(t)$. This lower bound quantifies connectivity in a way similar to the weighted Fiedler number. In this section we make this connection more precise and apply the spectral method to develop a conditional alignment criterion for singular local kernels of rather general nature: $\phi(x, y, t) = \phi(y, x, t)$ and satisfying

$$\lambda \frac{\mathbb{1}_{|x-y|<r_0}}{|x - y|^{n+\alpha}} \leq \phi(x, y, t) \leq \frac{\Lambda(t)}{|x - y|^{n+\alpha}}, \qquad 0 < \alpha < 2, \qquad (6.20)$$

here $\Lambda(t)$ is simply assumed finite for any $t \geq 0$. Since the compactness of embedding $H^s \hookrightarrow L^2$ with $s > 0$ is crucial to this discussion, as well as a uniform lower bound on the density, which is consistent with finiteness of mass only on a compact environment Ω, we restrict ourselves to the periodic domain $\Omega = \mathbb{T}^n$.

A quantitative expression of coercivity of the commutator (6.2) relies on lower bounds on the density. Precisely how large that lower bound should be in order to ensure alignment is investigated in the following proposition.

Theorem 6.2 *Let ϕ be a symmetric, local, singular kernel satisfying (6.20), and let $(\rho, \mathbf{u})$ be a global strong solution to (6.1). Assume that*

$$C \geq \rho(t, x) \geq \frac{c}{\sqrt{1+t}}, \qquad C > c > 0. \tag{6.21}$$

Then the solution aligns at an algebraic rate. Namely, there exist $\eta > 0$ such that

$$\int_{\mathbb{T}^d} |\mathbf{u}(t, x) - \bar{\mathbf{u}}|^2 \rho(t, x) \, dx \leq \frac{1}{2M \, t^\eta}. \tag{6.22}$$

Proof We consider the family of eigenvalue problems parameterized by time:

$$\int_{\mathbb{T}^n} \phi(x, y, t)(\mathbf{u}(x) - \mathbf{u}(y)) \, dm_t(y) = \kappa(t)\mathbf{u}(x), \qquad \mathbf{u} \in H^{\frac{\alpha}{2}}. \tag{6.23}$$

We seek the minimal eigenvalue, which is of course 0 corresponding to the constant eigenfunction. To remove this trivial solution, we restrict it to the time-dependent one-codimensional subspace

$$H_0^{\frac{\alpha}{2}} = \left\{ \mathbf{u} \in H^{\frac{\alpha}{2}} : \int_{\mathbb{T}^n} \mathbf{u} \, dm_t = 0 \right\}. \tag{6.24}$$

The key issue here is that $H_0^{\frac{\alpha}{2}}$ depends on time. We will return to this later. So, we seek the *second* minimal eigenvalue of (6.23) restricted to $H_0^{\frac{\alpha}{2}}$, as a solution to the variational problem:

$$\kappa_2(t) = 2M \inf_{\mathbf{u} \in H_0^{\frac{\alpha}{2}}} \frac{\int_{\mathbb{T}^{2n}} \phi(x, y, t)|\mathbf{u}(y) - \mathbf{u}(x)|^2 \rho(t, y)\rho(t, x) \, dx \, dy}{\int_{\mathbb{T}^{2n}} |\mathbf{u}(x) - \mathbf{u}(y)|^2 \rho(t, x)\rho(t, y) \, dx \, dy}. \tag{6.25}$$

In view of (6.20), and the assumed bounds on the density (6.21), the upper norm is equivalent to the norm of $H^{\alpha/2}$ and the lower to the norm of L^2. So, the existence follows classically by compactness. The number $\kappa_2(t)$ bears complete resemblance to the discrete weighted Fiedler number, discussed in Sect. 2.3. In terms of this Fiedler number, the energy equation (6.5) takes form

$$\frac{d}{dt} \mathcal{V}_2 \leq -\kappa_2(t)\mathcal{V}_2. \tag{6.26}$$

Consequently

$$V_2(t) \le V_2(0) \exp\left\{-\int_0^t \kappa_2(s)\, ds\right\}. \tag{6.27}$$

We will derive now the lower bound $\kappa_2(t) \ge c/(1+t)$ which clearly implies the statement of the proposition. Using the bounds on the density (6.21), the mean-zero condition on $\mathbf{u}$, and the lower bound of the kernel (6.20), we obtain

$$\int_{\mathbb{T}^{2n}} |\mathbf{u}(x) - \mathbf{u}(y)|^2 \rho(t,x)\rho(t,y)\, dx\, dy = 2M \int_{\mathbb{T}^n} |\mathbf{u}(x)|^2 \rho(t,x)\, dx \le C\|\mathbf{u}\|_2^2,$$

$$\int_{\mathbb{T}^{2n}} \phi(x,y,t)|\mathbf{u}(y) - \mathbf{u}(x)|^2 \rho(t,y)\rho(t,x)\, dx\, dy$$

$$\ge \frac{c}{t} \int_{\mathbb{T}^{2n}} \phi(x,y,t)|\mathbf{u}(y) - \mathbf{u}(x)|^2\, dx\, dy$$

$$\ge \frac{c}{t} \int_{|x-y|<r_0} \frac{|\mathbf{u}(x) - \mathbf{u}(y)|^2}{|x-y|^{n+\alpha}}\, dx\, dy,$$

and hence

$$\kappa_2(t) \ge \frac{c}{t} \inf_{\mathbf{u} \in H_0^{\frac{\alpha}{2}}} \frac{\int_{|x-y|<r_0} \frac{|\mathbf{u}(x)-\mathbf{u}(y)|^2}{|x-y|^{n+\alpha}}\, dx\, dy}{\|\mathbf{u}\|_2^2}. \tag{6.28}$$

Technically, the infimum still depends on time since the mean-zero condition is time dependent. So, the last piece to show is that this infimum stays bounded away from zero. We argue by contradiction. Suppose there is a sequence of times $t_k > 0$ and $\mathbf{u}_k \in H^{\alpha/2}$ with $\int \mathbf{u}_k\, dm_{t_k} = 0$ such that $\|\mathbf{u}_k\|_2 = 1$ and

$$\int_{|x-y|<r_0} \frac{|\mathbf{u}_k(x) - \mathbf{u}_k(y)|^2}{|x-y|^{n+\alpha}}\, dx\, dy \to 0. \tag{6.29}$$

The latter, in particular, implies compactness of the sequence $\{\mathbf{u}_k\}_k$ in L^2. Hence, up to a subsequence, $\mathbf{u}_k \to \mathbf{u}$ strongly in L^2 and weakly in $H^{\alpha/2}$. By the lower weak semi-continuity, and (6.29), we conclude that $\|\mathbf{u}\|_{\dot{H}^{\alpha/2}} = 0$ and hence $\mathbf{u}$ is a constant field, with $|\mathbf{u}| = 1$ due to $\|\mathbf{u}_k\|_2 \to \|\mathbf{u}\|_2$.

At the same time, since $\int \rho(t_k, x)\, dx = M$, there exists a weak* limit of a further subsequence $m_{t_k} \to m$, where m is a positive Radon measure on the torus $\mathbb{T}^n$ with nontrivial total mass $m(\mathbb{T}^n) = M$. We now reach a contradiction if we prove the limit:

$$0 = \int_{\mathbb{T}^n} \mathbf{u}_k(x)\rho(t_k, x)\, dx \to M\mathbf{u} \ne 0.$$

To prove the claimed limit, note that the assumed uniform upper bound on the density implies

$$\int_{\mathbb{T}^n} \mathbf{u}_k(x) \rho(t_k, x) \, dx - M\mathbf{u} = \int_{\mathbb{T}^n} (\mathbf{u}_k(x) - \mathbf{u}) \rho(t_k, x) \, dx,$$

and the latter is clearly bounded by $C\|\mathbf{u}_k - \mathbf{u}\|_2 \to 0$.

   This proves that $\kappa_2(t) \geq c/t$, and the result follows.   □

   In the course of the proof, we essentially established a statement analogous to Lemma 2.1 in the discrete case. Let us state it separately.

Lemma 6.1 *Let $\kappa_2(t)$ be the weighted Fiedler number defined by (6.25), and suppose that*

$$\int_0^\infty \kappa_2(s) \, ds = \infty.$$

Then the solution aligns: $\mathcal{V}_2 \to 0$.

   We can see now that unconditional flocking is generally achieved under the lower bound on the density, $\rho(t, \cdot) \gtrsim (1 + t)^{-1/2}$. The difficulty is that this lower bound is too restrictive and is not given a priori for any strong solution. The situation improves considerably for the topological models which yield unconditional flocking under more accessible assumptions on the density. We discuss those next.

6.4   Topological Models. Adaptive Diffusion

A field study of actual animal formations, such as flocks of starlings, see [5, 6, 20–23], confirmed that interactions in these biological system follow a somewhat different protocol. Each agent is capable to probe a certain fixed number of other agents in its proximity rather than all those within a predefined radius. This nearest neighbor rule defines a neighborhood determined by mass rather than Euclidean distance; see Fig. 6.2. Communication depending on the mass of the crowd is called *topological* as opposed to the conventional *metric* one based on the Euclidean distance $|\mathbf{x}_i - \mathbf{x}_j|$.

   In the context of Cucker-Smale flocking models, such topological communication can be defined using a topological distance which is given by the mass of a communication domain in the intermediate region between two agents $\Omega(\mathbf{x}_i, \mathbf{x}_j)$. For example

$$d(\mathbf{x}_i, \mathbf{x}_j) = \left[\sum_{k : \mathbf{x}_k \in \Omega(\mathbf{x}_i, \mathbf{x}_j)} m_k \right]^{\frac{1}{n}}, \tag{6.30}$$

Fig. 6.2 Topological versus metric communication

where the power $1/n$ is used to bring volume units back to length. Remarkably, models based on a topological protocol exhibit a much more robust flocking behavior than those based on a metric protocol, which we will demonstrate in this section.

We construct topological communication based on the following principles:

1. Every agent $\mathbf{x}_j$ has a finite range of influence, which is a Euclidean ball of radius r_0 centered at $\mathbf{x}_i$, denoted by $B(\mathbf{x}_i, r_0)$.
2. Agent $\mathbf{x}_i$ determines the distance to another agent $\mathbf{x}_j$ via $d(\mathbf{x}_i, \mathbf{x}_j)$, where $\Omega(\mathbf{x}_i, \mathbf{x}_j)$ is a symmetric communication domain, $\Omega(\mathbf{x}_i, \mathbf{x}_j) = \Omega(\mathbf{x}_j, \mathbf{x}_i)$, which includes the agents themselves $\mathbf{x}_i, \mathbf{x}_j \in \partial\Omega(\mathbf{x}_i, \mathbf{x}_j)$.
3. The communication ϕ_{ij} is inversely proportional to $d(\mathbf{x}_i, \mathbf{x}_j)$ based on the principle that in heavier regions information propagates slower.

Based on the outlined principles, we make the following choice:

$$\phi(\mathbf{x}_i, \mathbf{x}_j) = \frac{1}{d^{\tau}(\mathbf{x}_i, \mathbf{x}_j)} \psi(|\mathbf{x}_i - \mathbf{x}_j|), \tag{6.31}$$

where ψ is a nonnegative function with support containing the ball of radius r_0,

$$\psi(r) \geq \lambda \mathbb{1}_{r < r_0} \tag{6.32}$$

and $\tau > 0$ is a parameter. The kernel ψ encodes metric dependencies of the kernel, while τ gauges presence of topological effects.

The choice of the domain $\Omega(\mathbf{x}, \mathbf{y})$ can be rather flexible. The subsequent analysis goes through as long as it satisfies two basic requirements (refer to Fig. 6.3): (a) the region is a subset of the ball determined by $[\mathbf{x}, \mathbf{y}]$ as its diameter chord, and two cones of opening $< \pi$ at vertices $\mathbf{x}$ and $\mathbf{y}$, b) $\Omega(\mathbf{x}, \mathbf{y}) = \Omega(\mathbf{y}, \mathbf{x})$, and the boundary of the region is smooth everywhere except for $\mathbf{x}, \mathbf{y}$. For simplicity we also assume that topological communication is homogeneous and isotropic, i.e., $\Omega(\mathbf{x}, \mathbf{y})$ is constructed by a shift, rotation, and dilation of a basic domain $\Omega(-\mathbf{e}_1, \mathbf{e}_1)$. In order to insure symmetry, (b) we assume that $\Omega(-\mathbf{e}_1, \mathbf{e}_1) = -\Omega(-\mathbf{e}_1, \mathbf{e}_1)$.

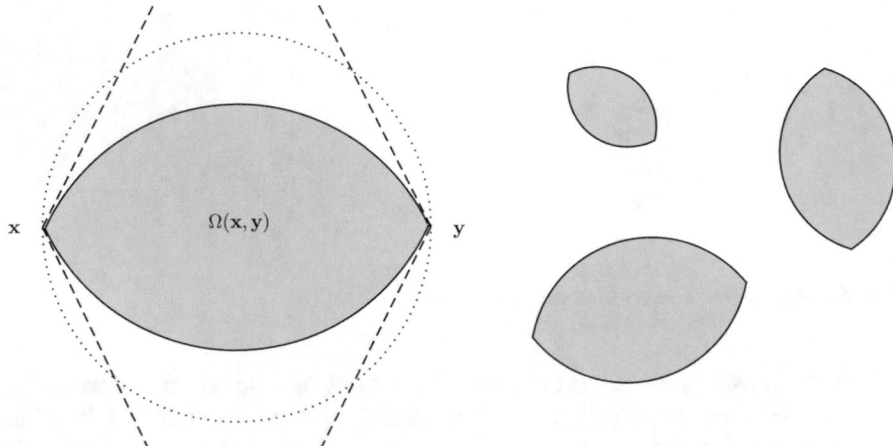

Fig. 6.3 Communication domains between agents

The analogue of Theorem 2.2 follows along the lines of the proof presented in Sect. 2.4. We have

$$\int_{\mathcal{D}_0}^{\infty} \psi(r)\, dr > \frac{\mathcal{A}_0}{\lambda M^{1-\frac{\tau}{n}}} \quad \Rightarrow \quad \mathcal{A}(t) \le \mathcal{A}_0 e^{-t\lambda M^{1-\frac{\tau}{n}}\psi(\overline{\mathcal{D}})}. \tag{6.33}$$

We are however mostly focused on local interactions. So, this observation applies only in special cases, for instance, when the initial flock is very aggregated $\mathcal{D}_0 \ll r_0$ and almost aligned $\mathcal{A}_0 \ll \lambda M^{1-\frac{\tau}{n}}$.

Note that in the macroscopic limit $N \to \infty$, the natural interpretation of the topological distance is given by

$$d(x, y) = \left[\int_{\Omega(x,y)} \rho(t, z)\, dz \right]^{\frac{1}{n}}.$$

In fact in 1D, where

$$d(x, y) = \left| \int_x^y \rho(t, z)\, dz \right|,$$

this does define a proper metric.[1] Otherwise, strictly speaking, d is a quasi-metric under the no-vacuum condition.

---

[1]In the notation for d and the kernel, we will omit the dependence on time for brevity.

The corresponding hydrodynamic topological system is given by

$$\begin{cases} \rho_t + \nabla \cdot (\rho \mathbf{u}) = 0, \\[2mm] \mathbf{u}_t + \mathbf{u} \cdot \nabla_x \mathbf{u} = \displaystyle\int_\Omega \phi(x, y)(\mathbf{u}(y) - \mathbf{u}(x))\rho(y)\,dy, \\[2mm] \phi(x, y) = \dfrac{\psi(|x - y|)}{d^\tau(x, y)}. \end{cases} \tag{6.34}$$

A proper care has to be given to properly define the singular integral operator $\mathcal{L}_\phi f$, and the commutator $\mathcal{C}_\phi$ is this case. These issues as well as well-posedness of (6.34) are discussed in [88, 97] and will be omitted here as they go beyond the scope of this text.

Note that the new kernel incorporates a type of *adaptive diffusion* which enhances dissipation into thinner regions and moderates it in thicker regions. As a result we can obtain an improvement upon Proposition 6.2 which requires a much weaker lower bound on the density, one that we will be able to actually prove at least in a 1D situation.

Theorem 6.3 *Let $(\rho, \mathbf{u})$ be a global smooth solution of the topological model (6.34) on the torus $\mathbb{T}^n$, with $\tau \geq n$. Assume that the density satisfies the lower bound:*

$$\rho(t, x) \geq \frac{c}{1 + t}, \qquad \forall t > 0, \ x \in \mathbb{T}^n. \tag{6.35}$$

Then the solution aligns with a logarithmic rate given by

$$\|\mathbf{u}(t) - \bar{\mathbf{u}}\|_\infty \leq \frac{c}{\sqrt{\ln t}}. \tag{6.36}$$

Remark 6.2 The assumption on the density (6.35) holds automatically in the 1D case, as we will see in Sect. 8; see (8.68).

Proof We aim to prove (6.36) for each component of the velocity u_i. Let us denote $u = u_i$ for short. By the Galilean invariance of the system, we can add a constant to u if necessary and assume that $u(t) > 0$. By the maximum principle, the extrema of $u(t)$, denoted $u_+(t)$ and $u_-(t)$, are monotone.

Step 1: Alignment Near Extremes Denote by $x_+(t)$ a point where the maximum of $u(t, \cdot)$ is achieved and by $x_-(t)$ for minimum. Let us fix a time-dependent $\delta(t) > 0$ to be specified later. Consider the sets

$$G_\delta^+(t) = \{u < u_+(t)(1 - \delta(t))\}, \qquad G_\delta^-(t) = \{u > u_-(t)(1 + \delta(t))\}.$$

The amount of flattening near extreme values will be quantified in terms of conditional expectations of the above sets relative to the local balls $B(x_\pm(t), r_0)$. We denote such expectations by

$$\mathbb{E}_t[A|B] = \frac{m_t(A \cap B)}{m_t(B)}.$$

First, let us show that

$$\int_0^\infty \delta(t)\mathbb{E}_t[G_\delta^\pm(t)|B(x_\pm(t), r_0)]\,dt < \infty. \tag{6.37}$$

To this end let us compute the equation at $(x_+(t), t)$ (with the use of Rademacher's Lemma 2.2):

$$\frac{d}{dt}u_+ = \int_{\mathbb{T}^n} \phi(x_+, y)(u(y) - u_+)\rho(y)\,dy.$$

We now use our assumption (6.32) and the fact that $\tau \geq n$ to obtain the bound:

$$\frac{\lambda \mathbb{1}_{r<r_0}(|x - y|)}{d^n(x, y)} \leq \frac{\psi(x - y)d^{\tau-n}(x, y)}{d^\tau(x, y)} \leq M^{\tau-n}\phi(x, y). \tag{6.38}$$

Thus, we have

$$-\frac{d}{dt}u_+ = \int \phi(x_+, y)(u_+ - u(y))\rho(y)\,dy$$

$$\gtrsim \int_{B(x_+,r_0)} \frac{1}{d(x_+, y)}(u_+(t) - u(y))\rho(y)\,dy$$

$$\geq \frac{1}{m_t(B(x_+, r_0))} \int_{G_\delta^+(t)\cap B(x_+,r_0)} (u_+ - u(y))\rho(y)\,dy$$

and since $\Omega(x_+, y) \subset B(x_+, r_0)$

$$\geq \frac{\delta(t)u_+}{m_t(B(x_+, r_0))} \int_{G_\delta^+(t)\cap B(x_+,r_0)} \rho(y)\,dy$$

$$= \delta(t)u_+\mathbb{E}_t[G_\delta^+(t)|B(x_+, r_0)].$$

Integrating we obtain

$$\int_0^\infty \delta(t)\mathbb{E}_t[G_\delta^+(t)|B(x_+(t), r_0)]\,dt \lesssim \ln \frac{u_+(0)}{\lim_{t\to\infty} u_+(t)} \leq \ln \frac{u_+(0)}{u_-(0)}.$$

Step 2: Use of Campanato-Morrey Norm In this step we show that u does not deviate much from its averages over local balls, whereby essentially establishing local alignment as depicted in Fig. 6.1. We express such measure of deviation in terms of a Campanato-Morrey metric to be specified later. Let us denote

Fig. 6.4 $\Omega(x, x')$ is trapped in the outer ball if x is close to the center x^*

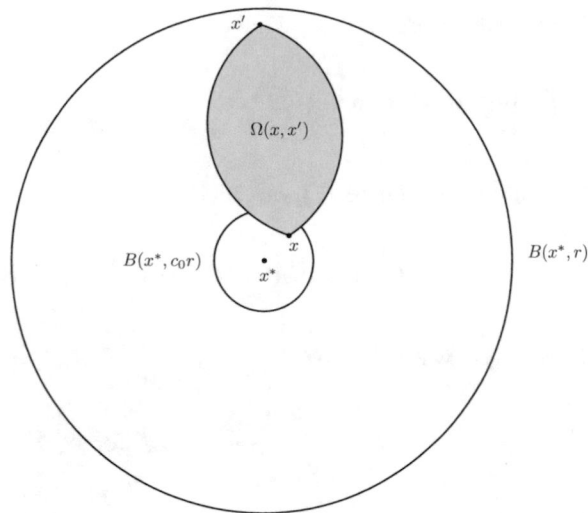

$$u_{x,r} = \frac{1}{m_t(B(x,r))} \int_{B(x,r)} u(t, z) \, dm_t(z).$$

Recall that the communication domains $\Omega(x, x')$ are confined to cones intersected with the diametric ball. As a result the following geometric observation is true; see Fig. 6.4.

Claim There exists a $c_0 > 0$ depending only on the opening angle of the cones such that for any $r > 0$ and any triple of points (x, x', x^*) with $|x - x^*| < c_0 r$ and $|x' - x^*| < r$, we have $\Omega(x, x') \subset B(x^*, r)$.   □

Let us fix an arbitrary $x^* \in \mathbb{T}^n$. By Hölder inequality, we have the following estimate for any $r < r_0/2$:

$$\int_{|x-x^*|<c_0 r} |u(x) - u_{x^*,r}|^2 \rho(x) \, dx$$

$$\leq \int_{\substack{|x-x^*|<c_0 r \\ |x'-x^*|<r}} \frac{1}{m_t(B(x^*,r))} |u(x) - u(x')|^2 \rho(x)\rho(x') \, dx' \, dx$$

using that $m_t(B(x_*, r)) \geq m_t(\Omega(x, x')) = d^n(x, x')$

$$\leq \int_{|x-x'|<(1+c_0)r} \frac{1}{d^n(x, x')} |u(x) - u(x')|^2 \rho(x)\rho(x') \, dx' \, dx$$

$$\leq C \int_{\mathbb{T}^{2n}} \phi(x, x') |u(x) - u(x')|^2 \rho(x)\rho(x') \, dx' \, dx.$$

From the energy equality (6.5), the right-hand side is globally integrable on $\mathbb{R}_+$. Consequently, we obtain a global time integrability for the following Campanato-

Morrey semi-norm

$$\int_0^\infty [u]_{r_0}^2 \, dt < \infty, \qquad [u]_{r_0}^2 := \sup_{x^* \in \mathbb{T}^n, r < \frac{r_0}{2}} \int_{|x-x^*|<c_0 r} |u(x) - u_{x^*,r}|^2 \rho(x) \, dx.$$

In combination with (6.37), we obtain

$$I = \int_0^\infty \left(\delta(t) \mathbb{E}_t [G_\delta^\pm(t) | B(x_\pm(t), r_0)] + [u(t)]_{r_0}^2 \right) dt < \infty.$$

Denoting $A = e^{2I}$ we have

$$\int_T^{T^A} \frac{dt}{t \ln t} = 2I \quad \text{for all } T > 0.$$

Consequently, for any $T > 0$, there exists a $t \in [T, T^A]$ such that

$$[u(t)]_{r_0}^2 < \frac{1}{t \ln t},$$

$$\mathbb{E}_t[G_\delta^+(t) | B(x_+(t), r_0)] + \mathbb{E}_t[G_\delta^-(t) | B(x_-(t), r_0)] < \frac{1}{\delta(t) t \ln t}. \tag{6.39}$$

By virtue of the lower bound on the density (6.35), this implies that

$$\sup_{x^*, r < \frac{r_0}{2}} \int_{|x-x^*|<c_0 r} |u(x) - u_{x^*,r}|^2 \, dx \le \frac{1}{\ln t}. \tag{6.40}$$

Step 3: Sliding Averages Let $t \in [T, T^A]$ be the moment of time fixed above. Let us fix $r = \frac{1}{4} r_0$, which lies within the reach of the Campanato metric. We will connect the two averages $u_{x_+,r}$ and $u_{x_-,r}$ sliding along the line connecting x_+ and x_- and show that the fluctuation of those averages is small.

To this end, consider the direction vector $\bar{n} = \frac{x_+ - x_-}{|x_+ - x_-|}$ and define a sequence of balls, $B_k = B(x_k, c_0 r)$, $k = 0, \dots, K$, with centers given by $x_0 = x_-$ and defined recursively by $x_{k+1} = x_k + c_0 r \bar{n}$ up to $k = K - 1$ and ending with $x_K = x_+$. The point is that the balls overlap significantly: $|B_k \cap B_{k+1}| \ge c_1 r_0^n$.

By the Chebyshev inequality, followed by (6.40) applied to the ball centered at x_0, we yield

$$|\{x \in B_0 \cap B_1 : |u(x) - u_{x_0,r}| > \eta\}| \le \frac{1}{\eta^2} \int_{B_0} |u(x) - u_{x_0,r}|^2 \, dx \le \frac{1}{\eta^2 \ln t}.$$

Let us set $\eta = \dfrac{2}{\sqrt{c_1 r_0^n \ln t}}$ so that

$$|\{x \in B_0 \cap B_1 : |u(x) - u_{x_0,r}| > \eta\}| \le \frac{1}{4}|B_0 \cap B_1|.$$

By the same argument applied to the fluctuation around the averaged value $u_{x_1,r}$, we obtain

$$|\{x \in B_0 \cap B_1 : |u(x) - u_{x_1,r}| > \eta\}| \le \frac{1}{4}|B_0 \cap B_1|.$$

Hence, the complements of the two sets must have a common point in the intersection $B_0 \cap B_1$:

$$\{x \in B_0 \cap B_1 : |u(x) - u_{x_0,r}| \le \eta\} \cap \{x \in B_0 \cap B_1 : |u(x) - u_{x_1,r}| \le \eta\} \ne \emptyset.$$

This implies

$$|u_{x_0,r} - u_{x_1,r}| \le 2\eta.$$

Continuing in a similar fashion, we recover the same bound for all consecutive pairs of averages:

$$|u_{x_k,r} - u_{x_{k+1},r}| \le 2\eta.$$

Hence

$$|u_{x_-,r} - u_{x_+,r}| \le 2K\eta \lesssim \frac{1}{\sqrt{\ln t}}. \tag{6.41}$$

Notice the absolute bound on $K \lesssim 1/r_0$. Furthermore, in view of (6.39), the following estimate holds

$$u_{x_+,r} \ge \frac{1}{m_t(B(x_+, r))} \int_{B(x_+,r)\setminus G_\delta^+} u_+(t)(1 - \delta(t))\, dm_t$$

$$\ge u_+(t)(1 - \delta(t))(1 - \mathbb{E}_t[G_\delta^+(t)|B(x_+(t), R_0)])$$

$$\ge u_+(t)(1 - \delta(t))\left(1 - \frac{1}{\delta(t)t\ln t}\right).$$

Consequently

$$u_+(t) - u_{x_+,r}(t) \lesssim \delta(t) + \frac{1}{\delta(t)t\ln t} \lesssim \frac{1}{\sqrt{t\ln t}},$$

provided we make the following choice of δ

$$\delta(t) = \frac{1}{\sqrt{t\ln t}}.$$

The same argument can be used to estimate the bottom average. In combination with (6.41), these imply

$$|u_+(t) - u_-(t)| \lesssim \frac{1}{\sqrt{\ln t}}.$$

To conclude we notice that the maximum principle implies

$$|u_+(T^A) - u_-(T^A)| \lesssim \frac{1}{\sqrt{\ln t}} \sim \frac{1}{\sqrt{\ln(T^A)}}.$$

Since T is arbitrary, the proof is complete.   □

6.5   Notes and References

The pressureless Euler alignment system was derived shortly after Cucker and Smale's seminal work [33] by Ha and Tadmor [45] who also established energy-based flocking behavior. Theorem 6.1 appeared in [99] almost in its present form, and the stability result is technically new; however it is a direct consequence of the analysis presented in the kinetic context.

The topological model discussed here was introduced in Shvydkoy and Tadmor [97] where Theorems 6.2 and 6.3 were proved. A prior prototypical model in the context of Cucker-Smale dynamics was proposed by Haskovec in [51]. The protocol involves mass around the agent $d(x, y) = m_t(B(x, |x - y|))$. This metric is embedded into the classical kernel (1.3), which makes it non-symmetric. Nonetheless alignment is established under an infinitely recurring in time connectivity assumption. A similar mass-distance was exploited in kinetic rank-based models by Blanchet and Degond in [7, 8]; see also [56]. Mixed metric-topological models appeared in [78, 89]. We remark that the metric component in our model also plays a crucial role in establishing a priori bounds on the density [97]. For further discussion see Sect. 8.5.

Theorem 6.3 allows for various extensions, which improve upon the rate of alignment under a more restrictive bound from below on the density. Specifically, the following statement can be proved along the lines of the argument presented here: suppose

$$\rho(t, x) \geq \frac{c}{(1 + t)^\gamma}, \quad 0 \leq \gamma \leq 1; \tag{6.42}$$

then the solution aligns with the following algebraic rate:

$$\|\mathbf{u}(t) - \bar{\mathbf{u}}\|_\infty = \frac{o(1)}{t^{\frac{1}{2}(1-\gamma)}}. \tag{6.43}$$

Chapter 7
Local Well-Posedness and Continuation Criteria

The regularity theory for models with smooth communication is understandably quite different from singular models—the former is essentially Burgers' equation with a damping mechanism, while the latter is a degenerate fractional parabolic system with dissipation in the momentum equation. In this chapter we will go through the first routine but very much essential step—proving local existence of classical solutions and setting a convenient functional framework for further development of the regularity theory for the Euler Alignment Systems. We are not seeking results in the sharpest spaces in order to keep the exposition simple. However, we do require our local solutions to have a certain level of regularity for various phenomena to remain classically verifiable, such as mass conservation, and existence of a characteristic flow-map. Thus, the results presented below will respect such requirements. We will also limit ourselves to the metric models, where the estimates are not excessively contaminated with density-dependent coefficients.

The technique of this chapter follows the standard energy method with somewhat more sophisticated adaptation in the singular communication case. The important byproduct will be continuation criteria which allow us to prove future global existence results by simply controlling a few lower order quantities; see Theorems 7.2 and 7.4.

As we start using various function spaces, we reserve a short notation for the L^p-norms, $\| \cdot \|_p$, $1 \leq p \leq \infty$, if the norm is understood over the entire domain in question. Norms in all other spaces X will be denoted by the standard $\| \cdot \|_X$.

7.1 Smooth Models

Let us assume throughout that ϕ is sufficiently smooth to take as many derivatives as necessary in the course of our arguments below. We also assume that $\phi = \phi(x - y)$

© Springer Nature Switzerland AG 2021
R. Shvydkoy, *Dynamics and Analysis of Alignment Models of Collective Behavior*,
Nečas Center Series, https://doi.org/10.1007/978-3-030-68147-0_7

is of convolution type and that the environment domain is $\mathbb{R}^n$. The exact same results
will carry over to $\mathbb{T}^n$ with slight modifications.

Using the commutator structure of the alignment term and (6.3), we write the
system (6.1) as

$$
\begin{cases}
\rho_t + \nabla \cdot (\rho \mathbf{u}) = 0, \\
\mathbf{u}_t + \mathbf{u} \cdot \nabla \mathbf{u} = \phi * (\rho \mathbf{u}) - \mathbf{u}\, \phi * \rho.
\end{cases}
\tag{7.1}
$$

Suppose we would like to prove local existence of solutions in the Sobolev class
$u \in H^m$, $\rho \in H^k \cap L^1_+$, where L^1_+ denotes the set of nonnegative functions in L^1.
Note that the Sobolev embedding does not guarantee that $\rho \in L^1_+$ if it is in a higher
Sobolev class, yet $\mathbf{u} \in L^\infty$ automatically if $m > \frac{n}{2}$. Both conditions are natural
quantities to include in the class as they are controlled by the dynamics a priori.

One can obtain local existence rather easily for a viscous regularization:

$$
\begin{cases}
\rho_t + \nabla \cdot (\rho \mathbf{u}) = \varepsilon \Delta \rho, \\
\mathbf{u}_t + \mathbf{u} \cdot \nabla \mathbf{u} = \phi * (\mathbf{u}\rho) - \mathbf{u}\, \phi * \rho + \varepsilon \Delta \mathbf{u}.
\end{cases}
\tag{7.2}
$$

Indeed, we are going to denote the grand quantity $Z = (\mathbf{u}, \rho)$ and consider the
equivalent mild formulation of (7.1):

$$
Z(t) = e^{\varepsilon t \Delta} Z_0 + \int_0^t e^{\varepsilon(t-s)\Delta} \mathcal{N}(Z(s))\, \mathrm{d}s,
$$

where $\mathcal{N}(Z)$ denotes all the nonlinear terms in (7.1). The proof proceeds by the
standard contractivity argument. Let us fix $Z_0 \in H^m \times (H^k \cap L^1_+)$ and consider the
map:

$$
\mathbb{F}[Z](t) = e^{\varepsilon t \Delta} Z_0 + \int_0^t e^{\varepsilon(t-s)\Delta} \mathcal{N}(Z(s))\, \mathrm{d}s.
$$

We need to show that for some small T, this map is a contraction on
$C([0, T); B_1(Z_0))$, where $B_1(Z_0)$ is the unit ball in $X = H^m \times (H^k \cap L^1)$
centered at Z_0. Let us prove invariance, while contractivity follows similarly. First,
by continuity of the heat semigroup

$$
\| Z_0 - e^{\varepsilon t \Delta} Z_0 \|_X \leq \frac{1}{2},
$$

for small t. To estimate the integral, we first recall the analyticity property of the
heat semigroup:

$$
\| \nabla e^{\varepsilon t \Delta} f \|_p \lesssim \frac{1}{\sqrt{\varepsilon t}} \| f \|_p, \quad 1 \leq p \leq \infty.
$$

So, by considering the ρ-component, we obtain

$$\left\| \int_0^t e^{\varepsilon(t-s)\Delta} \nabla \cdot (\rho \mathbf{u}) \, ds \right\|_1 \le \int_0^t \frac{1}{\sqrt{\varepsilon(t-s)}} \| \mathbf{u} \rho(s) \|_1 \, ds$$

$$\le \frac{t^{1/2}}{\varepsilon^{1/2}} \sup_{0 \le s \le t} (\| u(s) \|_\infty \| \rho(s) \|_1) \le \frac{T^{1/2}}{\varepsilon^{1/2}} (\| Z_0 \|_X^2 + 1) < \frac{1}{2},$$

for small T.

For H^k we argue similarly as for L^2. So, let ∂^k be a multi-index partial derivative of order k. We will use the product estimate:

$$\| \partial^k (\mathbf{u} \rho)(s) \|_2 \le \| \mathbf{u} \|_\infty \| \rho \|_{H^k} + \| \mathbf{u} \|_{H^k} \| \rho \|_\infty.$$

It is clear at this point that in order to close the estimates in X, we need to assume that $\frac{n}{2} < k \le m$, in which case

$$\| \partial^k (\mathbf{u} \rho)(s) \|_2 \le \| Z(s) \|_X^2.$$

So, by the same argument, we obtain

$$\left\| \partial^k \int_0^t e^{\varepsilon(t-s)\Delta} \nabla \cdot (\rho \mathbf{u}) \, ds \right\|_2 \le \int_0^t \frac{1}{\sqrt{\varepsilon(t-s)}} \| \partial^k (\mathbf{u} \rho)(s) \|_2 \, ds < \frac{1}{2}.$$

For the velocity there are two terms to handle. For the transport part, the L^2-estimate is straightforward, and

$$\left\| \partial^m \int_0^t e^{\varepsilon(t-s)\Delta} \mathbf{u} \cdot \nabla \mathbf{u} \, ds \right\|_2 \le \int_0^t \frac{1}{\sqrt{\varepsilon(t-s)}} \| \partial^{m-1} (\mathbf{u} \cdot \nabla \mathbf{u})(s) \|_2 \, ds$$

$$\le \int_0^t \frac{1}{\sqrt{\varepsilon(t-s)}} (\| \mathbf{u} \|_{H^{m-1}} \| \nabla \mathbf{u} \|_\infty + \| \mathbf{u} \|_{H^m} \| \mathbf{u} \|_\infty) \, ds < \frac{1}{4},$$

provided $m > \frac{n}{2} + 1$ to ensure embedding of $W^{1,\infty}$ into H^m. The commutator term $\phi * (\mathbf{u} \rho) - \mathbf{u} \phi * \rho$ is even easier to deal with, since the derivatives are absorbed by the kernel except when all fall on u, which results in the same estimate.

We have shown that $\mathbb{F} : C([0, T); B_1(Z_0)) \to C([0, T); B_1(Z_0))$ is a contraction, and so, we obtain a local solution on a time interval dependent on ε. Denoting T^* the maximal time of existence in $C([0, T); X)$, we show that T^* depends only on the X-norm of the initial condition. We do it by establishing a priori estimates that are independent of ε and which will allow us to pass to the limit of vanishing viscosity. So, the grand quantity we are trying to control is

$$Y_{m,k} = \| \mathbf{u} \|_{H^m}^2 + \| \rho \|_{H^k}^2 + \| \rho \|_1^2.$$

To start, we write the continuity equation as

$$\rho_t + \mathbf{u} \cdot \nabla\rho + (\nabla \cdot \mathbf{u})\rho = 0.$$

So, testing with $\partial^{2k}\rho$, we obtain

$$\frac{d}{dt}\|\rho\|^2_{\dot{H}^k} = \int (\nabla \cdot \mathbf{u})|\partial^k\rho|^2 \, dx - \int (\partial^k(\mathbf{u} \cdot \nabla\rho) - \mathbf{u} \cdot \nabla\partial^k\rho)\partial^k\rho \, dx$$

$$- \int \partial^k((\nabla \cdot \mathbf{u})\rho)\partial^k\rho \, dx - \varepsilon\|\rho\|^2_{H^{k+1}}.$$

We dismiss the last term. Recalling the classical commutator estimate

$$\|\partial^k(fg) - f\partial^k g\|_2 \le \|\nabla f\|_\infty\|g\|_{\dot{H}^{k-1}} + \|f\|_{\dot{H}^k}\|g\|_\infty, \tag{7.3}$$

we obtain

$$\frac{d}{dt}\|\rho\|^2_{\dot{H}^k} \le \|\nabla\mathbf{u}\|_\infty\|\rho\|^2_{\dot{H}^k} + \|\mathbf{u}\|_{\dot{H}^k}\|\rho\|_{\dot{H}^k}\|\nabla\rho\|_\infty + \|\mathbf{u}\|_{\dot{H}^{k+1}}\|\rho\|_{\dot{H}^k}\|\rho\|_\infty$$

$$\le C(\|\nabla\mathbf{u}\|_\infty + \|\nabla\rho\|_\infty + \|\rho\|_\infty)Y_{m,k},$$

provided $m \ge k + 1$. The L^2 norm of ρ obeys a similar estimate trivially, and the L^1-norm is conserved.

In the velocity equation, we apply the same commutator estimate for the material derivative part:

$$\int \partial^m(\mathbf{u} \cdot \nabla\mathbf{u})\partial^m\mathbf{u} \, dx = - \int \nabla \cdot \mathbf{u}|\partial^m\mathbf{u}|^2 \, dx$$

$$+ \int [\partial^m(\mathbf{u} \cdot \nabla\mathbf{u}) - (\mathbf{u} \cdot \nabla\partial^m\mathbf{u})]\partial^m\mathbf{u} \, dx \lesssim \|\nabla\mathbf{u}\|_\infty\|u\|^2_{\dot{H}^m}.$$

For the alignment term, we can put all the derivatives onto the kernel whenever possible, and the only term that is left out is $|\partial^m\mathbf{u}|^2 \phi * \rho$ with $\phi * \rho$ clearly bounded by $\|\phi\|_\infty M$—an a priori conserved quantity. So, we obtain

$$\frac{d}{dt}\|\mathbf{u}\|^2_{\dot{H}^m} \le (\|\nabla\mathbf{u}\|_\infty + C(\|\phi\|_{C^m}, M))\|\mathbf{u}\|^2_{\dot{H}^m}.$$

The similar bound for $\frac{d}{dt}\|\mathbf{u}\|^2_2$ is derived trivially. So, we obtain

$$\frac{d}{dt}\|\mathbf{u}\|^2_{H^m} \le (\|\nabla\mathbf{u}\|_\infty + C(\|\phi\|_{C^m}, M))\|\mathbf{u}\|^2_{H^m}. \tag{7.4}$$

It is important to note that this bound is independent of the higher norms of the density. Combining the two inequalities, we obtain

$$\frac{\mathrm{d}}{\mathrm{d}t} Y_{m,k} \le C(\|\nabla \mathbf{u}\|_\infty + \|\nabla \rho\|_\infty + \|\rho\|_\infty + C(\|\phi\|_{C^m}, M)) Y_{m,k}. \tag{7.5}$$

Of course $\|\nabla \mathbf{u}\|_\infty + \|\nabla \rho\|_\infty + \|\rho\|_\infty \le Y_{m,k}$ provided $k > \frac{n}{2} + 1$, which adds the last restriction on the exponents for the argument to work. So, if $m \ge k+1 > \frac{n}{2}+2$, then

$$\frac{\mathrm{d}}{\mathrm{d}t} Y_{m,k} \le C_1 Y_{m,k} + C_2 Y_{m,k}^2.$$

Solving the associated Riccati equation gives a uniform bound on the X-norm on a time interval inversely proportional to $\|Z_0\|_X$ but independent of ε. Thus, solutions to (7.2) with the same initial data exist on a common time interval $[0, T_0]$ where they are uniformly bounded in $C([0, T_0]; X)$.

Let us also note that keeping the dissipative terms in the estimates above also shows that

$$\varepsilon \int_0^{T_0} (\|\rho(s)\|_{H^{k+1}}^2 + \|\mathbf{u}(s)\|_{H^{m+1}}^2)\, \mathrm{d}s < C,$$

where C is independent of ε. Then

$$\|Z_t\|_{L^2} \le \|Z\|_X^2 + \varepsilon \|Z\|_{H^2} \le \|Z\|_X^2 + \varepsilon \|Z\|_{H^{k+1} \times H^{m+1}}.$$

So, $Z_t \in L^2([0, T_0]; L^2)$. Passing to a subsequence, we find a weak limit $Z_\varepsilon \to Z$ in $L^\infty([0, T_0]; X)$ and $(Z_\varepsilon)_t \to Z_t$ in $L^2([0, T_0]; L^2)$ (technically, a limit in L^1 may end up being a measure of bounded variation; however as a member of H^k, it is absolutely continuous, hence in L^1). Since $Z_t \in L^2([0, T_0]; L^2)$, Z is weakly continuous with values in L^2. Since L^2 is dense in H^{-m} and H^{-k}, this implies weak continuity $Z \in C_w([0, T_0]; H^m \times H^k)$. Strong continuity of the density follows from the equations themselves:

$$\|\rho_t\|_{L^1} \le \|\rho \nabla \mathbf{u}\|_1 + \|\mathbf{u} \nabla \rho\|_1 \le \|Z\|_X^2 < C.$$

Further regularity in time of Z follows from measuring smoothness of the system one level down and performing similar product estimates as above.

Having established local existence in X let us come back to (7.5), and notice that this solution can in fact be extended beyond T_0 if we know that

$$\int_0^{T_0} \|\nabla \mathbf{u}\|_\infty\, \mathrm{d}t < \infty. \tag{7.6}$$

Indeed, $\|\rho\|_\infty$ can be bounded by solving the continuity equation along characteristics:

$$\rho(X(t,\alpha),t) = \rho(\alpha,0)\exp\left\{-\int_0^t \nabla\cdot\mathbf{u}(X(s,\alpha),s)\,ds\right\}. \tag{7.7}$$

Using (7.4) we see that $\|\mathbf{u}\|_{H^m}^2$ is also bounded uniformly, and hence so is $\|\nabla^2\mathbf{u}\|_\infty$ since $m > \frac{n}{2} + 2$. Bootstrapping further by differentiating the continuity equation, we bound $\|\nabla\rho\|_\infty$ in a similar fashion.

Having this continuation criterion at hand, we can further improve the local existence result by establishing control over $\|\nabla\mathbf{u}\|_\infty$ directly. First, by the maximum principle, $\|\mathbf{u}(t)\|_\infty \leq \|\mathbf{u}_0\|_\infty$. Writing the equation for one component $\partial_i u_j$ we have

$$(\partial_t + \mathbf{u}\cdot\nabla)\partial_i u_j + \partial_i\mathbf{u}\cdot\nabla u_j = (\partial_i\phi)*(u_j\rho) - \partial_i u_j\phi*\rho - u_j(\partial_i\phi)*\rho. \tag{7.8}$$

Evaluating at the maximum and minimum and adding up over i, j, we obtain

$$\frac{d}{dt}\|\nabla\mathbf{u}\|_\infty \leq \|\nabla\mathbf{u}\|_\infty^2 + CM\|\mathbf{u}\|_\infty + M\|\nabla\mathbf{u}\|_\infty.$$

Hence, $\|\nabla\mathbf{u}\|_\infty$ is uniformly bounded a priori on a time interval depending only on $\|\nabla\mathbf{u}_0\|_\infty^{-1}$. So, the continuation criterion allows to extend our local solution to that time interval.

Using the transport structure of the momentum equation, we can further relax (7.6) to a condition on the divergence of $\mathbf{u}$, which will be extremely useful in the future. Let us recall that we have uniform bounds on the deformation tensor of the characteristic flow map given by (6.14). To translate this to a bound on $\|\nabla\mathbf{u}\|_\infty$ in Eulerian coordinates, we invert the flow map:

$$\nabla\mathbf{u}(x,t) = \nabla^{-\top}X(X^{-1}(x,t),t)\nabla\mathbf{v}(X^{-1}(x,t),t).$$

Using that

$$|\nabla^{-\top}X(\alpha,t)| \leq \frac{C_1}{|\det\nabla X(\alpha,t)|}e^{C_2 t}$$

we recall the Liouville formula for the Jacobian:

$$\det\nabla X(\alpha,t) = \exp\left\{\int_0^t \nabla\cdot\mathbf{u}(X(\alpha,t),t)\,dt\right\}.$$

Consequently, the condition

$$\int_0^{T_0}\inf_{x\in\mathbb{R}^n}\nabla\cdot\mathbf{u}(t,x)\,dt > -\infty \tag{7.9}$$

guarantees that the Jacobian does not vanish. This establishes a uniform bound on $\|\nabla \mathbf{u}(t)\|_\infty$ on the time interval $[0, T_0)$.

Let us record the obtained results in the following theorem.

Theorem 7.1 (Local Existence of Classical Solutions) *Suppose $m \geq k+1 > \frac{n}{2} + 2$ and $(u_0, \rho_0) \in H^m \times (H^k \cap L_+^1)$. Then there exists a time $T_0 = T_0(\|\nabla \mathbf{u}_0\|_\infty^{-1}, M)$ and a unique solution to (7.1) on the time interval $[0, T_0)$ in the class*

$$\mathbf{u} \in C_w([0, T_0); H^m) \cap \mathrm{Lip}([0, T_0); H^{m-1}),$$

$$\rho \in C_w([0, T_0); H^k \cap L_+^1) \cap \mathrm{Lip}([0, T_0); H^{k-1} \cap L_+^1), \tag{7.10}$$

satisfying the given initial condition. Moreover, any classical local solution on $[0, T_0)$ in the class (7.10) and satisfying (7.9) can be extended beyond T_0.

Theorem 7.1 can be used as a stepping stone to obtain solutions with less smoothness, especially for ρ, as long as the regularity of $\mathbf{u}$ permits to define a smooth characteristic flow map with sufficient compactness properties.

So, let us first assume that $\mathbf{u}_0 \in H^m$, $m > \frac{n}{2} + 1$ and $\rho_0 \in L_+^1$, the most basic assumption on the density. Mollifying the data $((\mathbf{u}_0)_\varepsilon, (\rho_0)_\varepsilon)$, due to Theorem 7.1, we obtain a family of local solutions on a common time interval $[0, T_0)$, since $H^m \subset W^{1,\infty}$ and $\|\nabla (\mathbf{u}_0)_\varepsilon\|_\infty \leq \|\nabla \mathbf{u}_0\|_\infty$. We also note that the estimate (7.4) holds for any integer m; hence $\mathbf{u}_\varepsilon \in L^\infty([0, T_0); H^m)$ uniformly. We also have $\partial_t \mathbf{u}_\varepsilon \in L^\infty([0, T_0); H^{m-1}) \subset L^\infty([0, T_0); L^\infty)$. Let us note that $H^m \subset W^{1+\delta,\infty}$ for some $\delta > 0$, and the embedding is compact on any bounded set, and of course $W^{1+\delta,\infty} \subset L^\infty$. So, the Aubin-Lions-Simon Lemma implies that the family is compact in $C([0, T_0); W^{1+\delta,\infty})$ on any bounded set. Passing to a subsequence, we obtain a weak limit $\mathbf{u}$ in $L^\infty([0, T_0); H^m)$ and strong in $C([0, T_0); W^{1+\delta,\infty})$ on any bounded set. The velocity also belongs to $C_w([0, T_0); H^m)$ as a consequence of the two memberships. Similarly, the family of flows X_ε belongs to $L^\infty([0, T_0); W^{1,\infty}) \cap \mathrm{Lip}([0, T_0); L^\infty)$, so it is compact in $C([0, T_0); C^\delta)$, $\delta < 1$, on any bounded set. We can thus claim strong uniform convergence of the flow maps as well. Considering the solution to the continuity equation

$$\rho_\varepsilon(X_\varepsilon(t, \alpha), t) = \rho_\varepsilon(\alpha, 0) \exp\left\{ -\int_0^t \nabla \cdot \mathbf{u}_\varepsilon(X_\varepsilon(s, \alpha), s)\, ds \right\}, \tag{7.11}$$

we can clearly pass to the strong limit in L^1 and the limit satisfies (7.7).

Note that the density essentially plays the role of a passive scalar. In particular, if $\rho_0 \in L_+^1 \cap L^\infty$ initially, then by formula (7.7), it will remain in the same class on the entire time interval.

The argument above can be elevated to any higher smoothness $H^m \times (L_+^1 \cap W^{k,\infty})$ as long as $m > \frac{n}{2} + k + 1$. It proceeds by differentiating the continuity equation k times and running the same compactness procedure. Weak continuity of the density in $L_+^1 \cap W^{k,\infty}$ follows from the established regularity properties of the velocity field and the corresponding formula for the solution of $\partial^k \rho$. The continuation criterion (7.9) remains valid in this setting as well.

Theorem 7.2 (Local Existence of Strong Solutions) *Suppose $m > \frac{n}{2} + k + 1$, $k = 0, 1, \ldots$, and $(u_0, \rho_0) \in H^m \times (L^1_+ \cap W^{k,\infty})$. Then there exists a time $T_0 = T_0(\|\nabla \mathbf{u}_0\|_\infty^{-1}, M)$ and a unique solution to (7.1) on the time interval $[0, T_0)$ in the class:*

$$(u, \rho) \in C_w([0, T_0); H^m \times (L^1_+ \cap W^{k,\infty})). \tag{7.12}$$

Moreover, any such solution satisfying (7.9) can be extended beyond T_0.

Lastly, let us discuss a small initial data result on the periodic domain $\mathbb{T}^n$, which in fact follows from the previous computations. Specifically, let us look back at the equation (7.8) and evaluate it at the maximum for each coordinate and note that

$$\mathcal{C}_{\partial_i \phi}(u_j, \rho) \leq \mathcal{A}(t) \|\nabla \phi\|_\infty M,$$

where $\mathcal{A}$ is the global amplitude of the velocity field. Now, on the compact domain $\mathbb{T}^n$, we have $\inf \phi = c_0 > 0$, so

$$-\partial_i u_j \phi * \rho \leq -c_0 M \|\nabla \mathbf{u}\|_\infty,$$

at a point of maximum. This implies

$$\frac{\mathrm{d}}{\mathrm{d}t} \|\nabla \mathbf{u}\|_\infty \leq \|\nabla \mathbf{u}\|_\infty^2 + \mathcal{A}(t) \|\nabla \phi\|_\infty M - c_0 M \|\nabla \mathbf{u}\|_\infty.$$

So, if initially $\mathcal{A}_0 < \varepsilon^2$ and $\|\nabla \mathbf{u}_0\|_\infty < \varepsilon$, then $\mathcal{A}(t) < \varepsilon^2$ for all times, and also for some short period of time $[0, t^*)$, we still have $\|\nabla \mathbf{u}(t)\|_\infty < 2\varepsilon$. Let t^* denote the maximal such time on the local interval of existence $[0, T)$. Then for all $t < t^*$, we have

$$\frac{\mathrm{d}}{\mathrm{d}t} \|\nabla \mathbf{u}\|_\infty \leq \varepsilon^2 \|\nabla \phi\|_\infty M - (c_0 M - \varepsilon) \|\nabla \mathbf{u}\|_\infty.$$

Integrating we obtain

$$\|\nabla \mathbf{u}(t)\|_\infty < \varepsilon + \frac{\varepsilon^2 \|\nabla \phi\|_\infty M}{c_0 M - \varepsilon}, \qquad t < t^*.$$

Provided $\varepsilon < \frac{c_0 M}{\|\nabla \phi\|_\infty M + 1}$, we have $\|\nabla \mathbf{u}\|_\infty < 2\varepsilon$. This implies that $t^* = T$, and, by the continuation criterion (7.9), $T = \infty$.

Theorem 7.3 (Small Initial Data) *Suppose (u_0, ρ_0) is in the class of spaces as stated in either Theorem 7.1 or Theorem 7.2 on the periodic domain $\mathbb{T}^n$. Suppose* *that*

$$\mathcal{A}_0 < \varepsilon^2, \quad \|\nabla \mathbf{u}_0\|_\infty < \varepsilon, \quad \varepsilon < \frac{c_0 M}{\|\nabla \phi\|_\infty M + 1}. \tag{7.13}$$

Then there exists a global in time solution in the same class for which one has

$$\|\nabla \mathbf{u}(t)\|_\infty < 2\varepsilon, \quad t > 0.$$

Let us note that on the open space this theorem would not hold. Indeed, we will see in Sect. 8 that if the quantity $\partial_x u_0 + \phi * \rho_0$ is negative at least at one point, then the solution blows up. So, no matter how small the initial slope is, one can always arrange a data with compact support of the density so that for some remote x we have $\partial_x u_0(x) = -\varepsilon$, yet $\phi * \rho_0(x) < \frac{1}{2}\varepsilon$. This solution will blow up in vacuum. However, one can still construct a proper theory of regular solutions in Lagrangian coordinates on the initial domain $\Omega_0 = \text{Supp}\,\rho_0$ in the Sobolev classes $(\mathbf{v}, \rho) \in H^m(\Omega_0) \times H^{m-1}(\Omega_0)$ in 1D; see Leslie [69] for details.

7.2 Singular Models

The analysis of singular models presented in this book will be carried out on the periodic domain $\mathbb{T}^n$. This choice is motivated by a technical reason—the no-vacuum condition $\rho \geq c_0 > 0$ necessary to retain uniform parabolicity of the alignment term is compatible with finite mass only on a compact domain.

We will consider singular communication kernels on $\mathbb{T}^n$ given by

$$\phi_\alpha(z) = \sum_{k \in \mathbb{Z}^n} \frac{1}{|z + 2\pi k|^{n+\alpha}}, \quad 0 < \alpha < 2. \tag{7.14}$$

The corresponding singular integral operator $\mathcal{L}_\phi$ given by (6.4) defines the classical fractional Laplacian:

$$\Lambda_\alpha f(x) = p.v. \int_{\mathbb{T}^n} (f(x+z) - f(x))\phi_\alpha(z)\,dz. \tag{7.15}$$

We sometimes use the alternative open space representation:

$$\Lambda_\alpha f(x) = p.v. \int_{\mathbb{R}^n} (f(x+z) - f(x))\frac{dz}{|z|^{n+\alpha}}, \tag{7.16}$$

where f is extended periodically on the entire $\mathbb{R}^n$.

The fractional Sobolev spaces, $W^{s,p}$, $1 < p < \infty$, $0 < s < 1$, can be defined as

$$\|f\|_{\dot{W}^{s,p}} = \|\Lambda_\alpha f\|_p, \qquad \|f\|_{W^{s,p}} = \|f\|_{\dot{W}^{s,p}} + \|f\|_p.$$

Higher-order spaces $W^{m+s,p}$ are defined by $\|\partial^m f\|_{W^{s,p}}$, etc. More often we will use the following intrinsic formulas for the fractional Sobolev norm

$$\|f\|_{\dot{W}^{s,p}}^p = \int_{\mathbb{T}^{2n}} |f(x+z) - f(x)|^p \phi_{sp}(z)\, dz\, dx$$

$$= \int_{\mathbb{T}^n \times \mathbb{R}^n} \frac{|f(x+z) - f(x)|^p}{|z|^{n+sp}}\, dz\, dx.$$

We refer the reader to [101] for a comprehensive treatise on function spaces.

For brevity we will use the following notation for increments:

$$\partial_z f(x) = f(x+z) - f(x).$$

All our well-posedness results also hold for local communication kernels with a smooth cutoff, $\phi(r) = \phi_\alpha(r)h(r)$. The important fact is that the action of the corresponding singular integral operator $\mathcal{L}_\phi$ is coercive:

$$c_1\|f\|_{\dot{H}^\alpha} - c_2\|f\|_2 \leq \|\mathcal{L}_\phi f\|_2 \leq c_3\|f\|_{\dot{H}^\alpha} + c_4\|f\|_2. \tag{7.17}$$

Necessity of the no-vacuum condition can be easily seen by the following example in 1D. Let us consider a local kernel for simplicity, $\mathrm{Supp}\,\phi \subset B_1(0)$. Let the initial density be confined to a small ball, say $\mathrm{Supp}\,\rho_0 \subset B_\varepsilon(0)$, while $u_0 = 1$ on $B_{10}(0)$, $v_0 = 0$ on $B_{10+\varepsilon}(0)$ and smooth in between. Then the density will remain in $B_2(0)$ for a time period of at least $t < 1$, due to $u \leq 1$. During this time the momentum equation will remain a pure Burgers equation; hence the solution will evolve into a shock at a time $t \sim \varepsilon < 1$.

A more subtle blowup can be constructed even when the density vanishes at one point.

Example 7.1 We will see in Sect. 8 that in dimension 1 there is an extra conservation law for the quantity $u_x + \mathcal{L}_\phi \rho$. So, if initially zero, it stays zero for all time. Consider the full fractional Laplacian kernel with $\alpha = 1$, $\Lambda = \Lambda_1$. Then $\Lambda\rho = -u_x$, where $\Lambda = -(-\Delta)^{1/2}$. Hence, the whole system reduces to

$$\rho_t + (\rho H\rho)_x = 0, \tag{7.18}$$

where H is the Hilbert transform. This equation appeared in many different contexts; see [4] for a full account. To obtain a blowup, we assume that initially $\rho_0(x) = \rho_0(-x)$, $\rho_0(0) = \rho_0'(0) = 0$. It is easy to see that these conditions are preserved in time. Then, writing the equation for $\Lambda\rho$, we obtain

$$(\Lambda\rho)_t - H(\rho H\rho)_{xx} = 0.$$

Using the identity $H(\rho H\rho) = \frac{1}{2}(H\rho)^2 - \frac{1}{2}\rho^2$, we write

$$(\Lambda\rho)_t - (\Lambda\rho)^2 - H\rho H\rho_{xx} + \rho\rho_{xx} + (\rho_x)^2 = 0.$$

Evaluating at 0 we obtain the Riccati equation

$$(\Lambda\rho)_t = (\Lambda\rho)^2.$$

Note that $\Lambda\rho_0(0) > 0$. Hence, the solution blows up.

The goal of this section is to obtain a local well-posedness result stated next.

Theorem 7.4 (Local Existence of Classical Solutions) *Suppose $m > \frac{n}{2} + 1$, $0 < \alpha < 2$ and*

$$(\mathbf{u}_0, \rho_0) \in H^{m+1}(\mathbb{T}^n) \times H^{m+\alpha}(\mathbb{T}^n),$$

and $\rho_0(x) > 0$ for all $x \in \mathbb{T}^n$. Then there exists a time $T_0 > 0$ and a unique non-vacuous solution to (6.1) on the time interval $[0, T_0)$ in the class:

$$\begin{aligned}
\mathbf{u} &\in C_w([0, T_0); H^{m+1}) \cap L^2([0, T_0); \dot{H}^{m+1+\alpha/2}), \\
\rho &\in C_w([0, T_0); H^{m+\alpha}).
\end{aligned} \tag{7.19}$$

Moreover, any such solution satisfying

$$\sup_{t \in [0,T_0)} (\|\nabla\rho(t)\|_\infty + \|\nabla\mathbf{u}(t)\|_\infty) < \infty \tag{7.20}$$

can be extended beyond T_0.

Performing energy estimates in the same fashion as for smooth models will inevitably create a derivative overload on the density. Instead we consider another "almost conserved" quantity

$$e = \nabla \cdot \mathbf{u} + \mathcal{L}_\phi\rho, \tag{7.21}$$

which satisfies the equation

$$e_t + \nabla \cdot (\mathbf{u}e) = (\nabla \cdot \mathbf{u})^2 - \mathrm{Tr}(\nabla\mathbf{u})^2. \tag{7.22}$$

Let us derive it in general for the sake of completeness. Since ϕ is a convolution kernel, we have

$$\partial_t \mathcal{L}_\phi + \nabla \cdot \mathcal{L}_\phi(\rho\mathbf{u}) = 0. \tag{7.23}$$

Taking the divergence of the velocity equation, we obtain

$$\partial_t (\nabla \cdot \mathbf{u}) + \nabla \cdot [\mathbf{u} \cdot \nabla \mathbf{u}] = \nabla \cdot \mathcal{L}_\phi(\rho \mathbf{u}) - \nabla \cdot [\mathbf{u}\mathcal{L}_\phi \rho] \tag{7.24}$$

with

$$\nabla \cdot [\mathbf{u}\mathcal{L}_\phi \rho] = \mathcal{L}_\phi \rho \nabla \cdot \mathbf{u} + \mathbf{u} \cdot \nabla \mathcal{L}_\phi \rho,$$

and

$$\nabla \cdot [\mathbf{u} \cdot \nabla \mathbf{u}] = \mathrm{Tr}(\nabla \mathbf{u})^2 + \mathbf{u} \cdot \nabla(\nabla \cdot \mathbf{u}).$$

On the one hand, combining (7.23) and (7.24), we obtain

$$\partial_t e + \mathcal{L}_\phi \rho \nabla \cdot \mathbf{u} + \mathbf{u} \cdot \nabla e + \mathrm{Tr}(\nabla \mathbf{u})^2 = 0. \tag{7.25}$$

Adding and subtracting now $(\nabla \cdot \mathbf{u})^2$ produces (7.22).

From the order of terms that enter into the formula for e, it is clear that the natural correspondence in regularity for the state variables is

$$(\mathbf{u} \in H^{m+1}) \sim (\rho \in H^{m+\alpha}).$$

Note that in 1D the right-hand side vanishes and we have a perfect conservation law. This case will be discussed at length in Sect. 8.

The grand quantity to be estimated is

$$Y_m = \|\mathbf{u}\|_{H^{m+1}}^2 + \|e\|_{H^m}^2 + \|e\|_\infty + \|\rho\|_1 + \|\rho^{-1}\|_\infty,$$

which is equivalent to $Y_m \sim \|\mathbf{u}\|_{H^{m+1}}^2 + \|\rho\|_{H^{m+\alpha}}^2 + \|\rho^{-1}\|_\infty$ in view of (7.17).

Our strategy will be very similar to the smooth case, where we obtain local solutions via viscous regularization and prove a continuation criterion via a priori estimates on Y_m. We assume throughout that $m > \frac{n}{2} + 1$ and $0 < \alpha < 2$.

So, let us start with (7.2) and consider the mild formulation:

$$\rho(t) = e^{\varepsilon t \Delta} \rho_0 - \int_0^t e^{\varepsilon(t-s)\Delta} \nabla \cdot (\mathbf{u}\rho)(s) \, \mathrm{d}s,$$

$$u(t) = e^{\varepsilon t \Delta} \mathbf{u}_0 - \int_0^t e^{\varepsilon(t-s)\Delta} \mathbf{u} \cdot \nabla \mathbf{u}(s) \, \mathrm{d}s \tag{7.26}$$

$$+ \int_0^t e^{\varepsilon(t-s)\Delta} \mathcal{C}_\phi(\mathbf{u}, \rho)(s) \, \mathrm{d}s.$$

Let us denote as before

$$Z = (\rho, \mathbf{u}) \in X := H^{m+\alpha} \times H^{m+1},$$

and by $\mathbb{F}[Z](t)$ the right-hand side of the mild formulation (7.26). In order to apply the standard fixed point argument, we have to show that $\mathbb{F}$ leaves the set $C([0, T_{\delta,\varepsilon}); B_\delta(Z_0))$ invariant and that it is a contraction. Here $B_\delta(Z_0)$ is the ball in X of radius ε centered around the initial condition Z_0. We limit ourselves to showing details for invariance as the estimates involved in proving contractivity are similar.

First we assume that ρ has no vacuum: $\rho_0(x) \geq c_0 > 0$. Since the metric we are using for $\rho \in H^{m+\alpha}$ controls the L^∞ norm, if $\delta > 0$ is small enough, then for any $\|\rho - \rho_0\|_{H^{m+\alpha}} < \delta$, one obtains

$$\rho(x) > \frac{1}{2}c_0. \tag{7.27}$$

So, let us assume that $Z \in C([0, T); B_\delta(Z_0))$. It is clear that $\|e^{\varepsilon t \Delta}Z_0 - Z_0\|_X < \frac{\delta}{2}$ provided the time t is short enough. Then Z has some bound $\|Z\| \leq C$. Using this let us estimate the norms under the integrals. First, recall that $\|\Lambda_\alpha e^{\varepsilon t \Delta}\|_{L^2 \to L^2} \lesssim \frac{1}{(\varepsilon t)^{\alpha/2}}$. In the case $\alpha \geq 1$, we have

$$\left\| \partial^m \Lambda_\alpha \int_0^t e^{\varepsilon(t-s)\Delta} \nabla \cdot (\mathbf{u}\rho)(s)\,ds \right\|_2 \lesssim \int_0^t \frac{1}{(t-s)^{\alpha/2}} \|\partial^{m+1}(\mathbf{u}\rho)(s)\|_2\,ds$$

$$\leq \int_0^t \frac{1}{(t-s)^{\alpha/2}} \|\mathbf{u}\|_{\dot{H}^{m+1}} \|\rho\|_{\dot{H}^{m+\alpha}}\,ds \leq C^2 t^{1-\alpha/2} < \frac{\delta}{2},$$

provided $T = T(\delta, \varepsilon)$ is small enough. In the case $\alpha < 1$, we combine instead one full derivative with the heat semigroup, and the rest $\partial^{m+\alpha}$ gets applied to $\mathbf{u}\rho$, which produces a similar bound.

Moving on to the u-equation, we have

$$\left\| \partial^{m+1} \int_0^t e^{\varepsilon(t-s)\Delta} \mathbf{u} \cdot \nabla \mathbf{u}(s)\,ds \right\|_2 \lesssim \int_0^t \frac{1}{(t-s)^{1/2}} \|\partial^m(\mathbf{u} \cdot \nabla \mathbf{u})(s)\|_2\,ds$$

$$\leq \int_0^t \frac{1}{(t-s)^{\alpha/2}} \|\mathbf{u}\|_{\dot{H}^{m+1}} \|\mathbf{u}\|_{\dot{H}^m}\,ds \leq C^2 t^{1/2} < \frac{\delta}{4}.$$

As to the commutator form, for $\alpha \leq 1$, the computation is very similar: We combine one derivative with the heat semigroup, and for the rest we use (7.17)

$$\|\partial^m C_\phi(\mathbf{u}, \rho)\|_2 \lesssim \|\mathbf{u}\|_{\dot{H}^{m+1}} \|\rho\|_{H^{m+\alpha}} < C^2,$$

and the rest follows as before. When $\alpha > 1$, we combine α derivatives with the semigroup, and the rest follows as before.

We have proved that $\|\mathbb{F}[Z](t) - Z_0\|_X < \delta$, for a short time, and hence, $\mathbb{F}$ leaves $C([0, T(\delta, \varepsilon)); B_\delta(Z_0))$ invariant.

Now let us derive a priori estimates for viscous solutions independent of ε. Note that the dissipation terms in all the following computations are negative and as such will be ignored.

First, evaluating the continuity equation at a point of minimum x_- and denoting $\rho_- = \min \rho$, we readily obtain

$$\frac{d}{dt}\rho_- = -\rho_- \nabla \mathbf{u} + \varepsilon \Delta \rho(x_-) \geq -\rho_- \|\nabla \mathbf{u}\|_\infty.$$

Hence

$$\frac{d}{dt}\|\rho^{-1}\|_\infty \leq \|\rho^{-1}\|_\infty \|\nabla \mathbf{u}\|_\infty \leq \|\nabla \mathbf{u}\|_\infty Y_m.$$

Furthermore

$$\frac{d}{dt}\|e\|_\infty \leq \|\nabla \mathbf{u}\|_\infty \|e\|_\infty + \|\nabla \mathbf{u}\|_\infty^2 \leq \|\nabla \mathbf{u}\|_\infty Y_m. \tag{7.28}$$

Let us continue with estimates on the e-quantity. We have

$$\frac{d}{dt}\|e\|_{\dot{H}^m}^2 \leq \int_{\mathbb{T}^n} \left(\partial^m e \mathbf{u} \cdot \nabla \partial^m e + \partial^m e [\partial^m (\mathbf{u} \cdot \nabla e) - \mathbf{u} \cdot \nabla \partial^m e] \right.$$
$$\left. + \partial^m e \partial^m (e \nabla \cdot \mathbf{u}) + \partial^m e [(\nabla \cdot \mathbf{u})^2 - \text{Tr}(\nabla \mathbf{u})^2] \right) dx.$$

In the first term, we integrate by part and estimate

$$\left| \int_{\mathbb{T}^n} \partial^m e \, \mathbf{u} \cdot \nabla \partial^m e \, dx \right| \leq \|e\|_{\dot{H}^m}^2 \|\nabla \mathbf{u}\|_\infty.$$

For the next commutator term, we use (7.3)

$$\left| \int_{\mathbb{T}^n} \partial^m e [\partial^m (\mathbf{u} \cdot \nabla e) - \mathbf{u} \cdot \nabla \partial^m e] \, dx \right| \leq \|e\|_{\dot{H}^m}^2 \|\nabla \mathbf{u}\|_\infty + \|e\|_{\dot{H}^m} \|\mathbf{u}\|_{\dot{H}^m} \|\nabla e\|_\infty.$$

Using the Gagliardo-Nirenberg inequality, we estimate the latter term as

$$\|e\|_{\dot{H}^m} \|\mathbf{u}\|_{\dot{H}^m} \|\nabla e\|_\infty \leq \|e\|_{\dot{H}^m} \|\mathbf{u}\|_{\dot{H}^{m+1}}^{\theta_1} \|\nabla \mathbf{u}\|_\infty^{1-\theta_1} \|e\|_{\dot{H}^m}^{\theta_2} \|e\|_\infty^{1-\theta_2},$$

where $\theta_1 = \frac{n-2(m-1)}{n-2m}$ and $\theta_2 = \frac{2}{2m-n}$. The two exponents add up to 1, so by the generalized Young inequality

$$\leq (\|e\|_{\dot{H}^m}^2 + \|\mathbf{u}\|_{\dot{H}^{m+1}}^2)(\|e\|_\infty + \|\nabla \mathbf{u}\|_\infty) \leq (\|e\|_\infty + \|\nabla \mathbf{u}\|_\infty) Y_m.$$

The next term in the e-equation is estimated by the product formula

$$\|\partial^m (fg)\|_2 \leq \|f\|_{H^m} \|g\|_\infty + \|f\|_\infty \|g\|_{H^m}. \tag{7.29}$$

So, we have

$$\left| \int_{\mathbb{T}^n} \partial^m e \, \partial^m (e\nabla \cdot \mathbf{u}) \, dx \right| \leq \|e\|_{\dot{H}^m}^2 \|\nabla \mathbf{u}\|_\infty + \|e\|_{\dot{H}^m} \|e\|_\infty \|\mathbf{u}\|_{\dot{H}^{m+1}}$$

$$\leq (\|e\|_\infty + \|\nabla \mathbf{u}\|_\infty) Y_m.$$

Finally

$$\left| \int_{\mathbb{T}^n} \partial^m e \, [(\nabla \cdot \mathbf{u})^2 - \text{Tr}(\nabla \mathbf{u})^2] \, dx \right| \leq \|e\|_{\dot{H}^m} \|\mathbf{u}\|_{\dot{H}^{m+1}} \|\nabla \mathbf{u}\|_\infty \leq \|\nabla \mathbf{u}\|_\infty Y_m.$$

Thus

$$\frac{d}{dt} \|e\|_{\dot{H}^m}^2 \leq (\|e\|_\infty + \|\nabla \mathbf{u}\|_\infty) Y_m. \tag{7.30}$$

Next we perform the main technical estimate on the velocity equation. We have

$$\frac{d}{dt} \|\mathbf{u}\|_{\dot{H}^{m+1}}^2 = -\int_{\mathbb{T}^n} \partial^{m+1} (\mathbf{u} \cdot \nabla \mathbf{u}) \cdot \partial^{m+1} \mathbf{u} \, dx + \int_{\mathbb{T}^n} \partial^{m+1} C_\phi(\mathbf{u}, \rho) \cdot \partial^{m+1} \mathbf{u} \, dx.$$

The transport term is estimated using the classical commutator estimate

$$\int_{\mathbb{T}^n} \partial^{m+1} (\mathbf{u} \cdot \nabla \mathbf{u}) \cdot \partial^{m+1} \mathbf{u} \, dx = \int_{\mathbb{T}^n} \mathbf{u} \cdot \nabla (\partial^{m+1} \mathbf{u}) \cdot \partial^{m+1} \mathbf{u} \, dx$$

$$+ \int_{\mathbb{T}^n} [\partial^{m+1}, \mathbf{u}] \nabla \mathbf{u} \cdot \partial^{m+1} \mathbf{u} \, dx.$$

Then

$$\int_{\mathbb{T}^n} \mathbf{u} \cdot \nabla (\partial^{m+1} \mathbf{u}) \cdot \partial^{m+1} \mathbf{u} \, dx = -\int_{\mathbb{T}^n} \frac{1}{2} (\nabla \cdot \mathbf{u}) |\partial^{m+1} \mathbf{u}|^2 \, dx \leq \|\nabla \mathbf{u}\|_\infty \|\mathbf{u}\|_{\dot{H}^{m+1}}^2,$$

and using (7.3) we obtain

$$\int_{\mathbb{T}^n} [\partial^{m+1}, \mathbf{u}] \nabla \mathbf{u} \cdot \partial^{m+1} \mathbf{u} \, dx \leq \|\nabla \mathbf{u}\|_\infty \|\mathbf{u}\|_{\dot{H}^{m+1}}^2.$$

Thus,

$$\frac{d}{dt} \|\mathbf{u}\|_{\dot{H}^{m+1}}^2 \leq \|\nabla \mathbf{u}\|_\infty Y_m + \partial^{m+1} C_\phi(\mathbf{u}, \rho) \cdot \partial^{m+1} u.$$

Let us expand the commutator:

$$\partial^{m+1}\mathcal{C}_\phi(\mathbf{u}, \rho) = \sum_{l=0}^{m+1} \binom{m+1}{l} \mathcal{C}_\phi(\partial^l \mathbf{u}, \partial^{m+1-l}\rho).$$

One end-point case, $l = m + 1$, gives rise to a dissipative term:

$$\int_{\mathbb{T}^n} \mathcal{C}_\phi(\partial^{m+1}\mathbf{u}, \rho) \cdot \partial^{m+1}\mathbf{u}\,dx = -\frac{1}{2}\int_{\mathbb{T}^{2n}} \phi(z)|\delta_z\partial^{m+1}\mathbf{u}(x)|^2\rho(x+z)\,dz\,dx$$
$$-\frac{1}{2}\int_{\mathbb{T}^{2n}} \phi(z)\delta_z\partial^{m+1}\mathbf{u}(x)\partial^{m+1}\mathbf{u}(x)\delta_z\rho(x)\,dz\,dx.$$

The first term is bounded by

$$-\rho_-\int_{\mathbb{T}^{2n}} \phi(z)|\delta_z\partial^{m+1}\mathbf{u}(x)|^2\,dz\,dx \sim -\rho_-\|\mathbf{u}\|^2_{\dot{H}^{m+1+\frac{\alpha}{2}}},$$

which is the main dissipation term. The second is estimated as follows. Let us pick an $\varepsilon > 0$ so small that $1 + \frac{\alpha}{2} > \alpha + \varepsilon$. Then

$$\left|\int_{\mathbb{T}^{2n}} \phi(z)\delta_z\partial^{m+1}\mathbf{u}(x)\partial^{m+1}\mathbf{u}(x)\delta_z\rho(x)\,dz\,dx\right|$$
$$\leq \|\nabla\rho\|_\infty \int_{\mathbb{T}^n \times \mathbb{R}^n} \frac{|\partial^{m+1}\delta_z\mathbf{u}(x)|}{|z|^{n/2+\alpha-1+\varepsilon}}\frac{|\partial^{m+1}\mathbf{u}(x)|}{|z|^{n/2-\varepsilon}}\,dz\,dx$$
$$\leq \|\nabla\rho\|_\infty\|\mathbf{u}\|_{H^{m+1}}\|\mathbf{u}\|_{H^{m+\alpha+\varepsilon}} \leq \|\nabla\rho\|_\infty\|\mathbf{u}\|_{H^{m+1}}\|\mathbf{u}\|_{H^{m+1+\alpha/2}}$$
$$\leq \frac{1}{2}\rho_-\|\mathbf{u}\|^2_{H^{m+1+\alpha/2}} + \rho_-^{-1}\|\nabla\rho\|^2_\infty Y_m,$$

where the first term is absorbed into the dissipation term. So

$$\int_{\mathbb{T}^n} \mathcal{C}_\phi(\partial^{m+1}\mathbf{u}, \rho) \cdot \partial^{m+1}\mathbf{u}\,dx \lesssim -\rho_-\|\mathbf{u}\|^2_{\dot{H}^{m+1+\frac{\alpha}{2}}} + \rho_-^{-1}\|\nabla\rho\|^2_\infty Y_m.$$

Let us consider first the other end-point case of $l = 0$. In this case the density suffers a derivative overload. We apply the following "easing" formula:

$$\int_{\mathbb{T}^n} \mathcal{C}_\phi(\mathbf{u}, \partial^{m+1}\rho) \cdot \partial^{m+1}\mathbf{u}\,dx = \int_{\mathbb{T}^{2n}} \phi(z)\delta_z\mathbf{u}(x)\partial^{m+1}\rho(x+z)\partial^{m+1}\mathbf{u}(x)\,dz\,dx.$$

Observe that

$$\partial^{m+1}\rho(x+z) = \partial_z\partial_x^m\rho(x+z) = \partial_z(\partial_x^m\rho(x+z) - \partial_x^m\rho(x)) = \partial_z\delta_z\partial^m\rho(x).$$

Let us now integrate by parts in z:

$$\int_{\mathbb{T}^n} C_\phi(\mathbf{u}, \partial^{m+1}\rho) \cdot \partial^{m+1}\mathbf{u}\, dx = \int_{\mathbb{T}^{2n}} \partial_z \phi(z) \delta_z \mathbf{u}(x) \delta_z \partial^m \rho(x) \partial^{m+1}\mathbf{u}(x)\, dz\, dx$$

$$+ \int_{\mathbb{T}^{2n}} \phi(z) \partial \mathbf{u}(x+z) \delta_z \partial^m \rho(x) \partial^{m+1}\mathbf{u}(x)\, dz\, dx := J_1 + J_2.$$

Let us start with J_2 first. By symmetrization

$$J_2 = \int_{\mathbb{T}^{2n}} \delta_z \partial \mathbf{u}(x) \delta_z \partial^m \rho(x) \partial^{m+1}\mathbf{u}(x) \phi(z)\, dz\, dx$$

$$- \int_{\mathbb{T}^{2n}} \partial \mathbf{u}(x) \delta_z \partial^m \rho(x) \delta_z \partial^{m+1}\mathbf{u}(x) \phi(z)\, dz\, dx$$

$$:= J_{2,1} + J_{2,2}.$$

The term $J_{2,1}$ will appear in a series of similar terms that we will estimate systematically below. The bound for $J_{2,2}$ is rather elementary:

$$J_{2,2} \leq \|\nabla \mathbf{u}\|_\infty \|\mathbf{u}\|_{\dot{H}^{m+1+\alpha/2}} + \|\rho\|_{H^{m+\alpha/2}} \leq \varepsilon \rho_- \|\mathbf{u}\|_{\dot{H}^{m+1+\alpha/2}}^2 + \rho_-^{-1} \|\nabla \mathbf{u}\|_\infty^2 Y_m.$$

A similar computation can be made for J_1. Indeed, using that $\partial_z \phi(z)$ is odd, by symmetrization, we have

$$J_1 = \frac{1}{2} \int_{\mathbb{T}^{2n}} \partial_z \phi(z) \delta_z \mathbf{u}(x) \delta_z \partial^m \rho(x) \delta_z \partial^{m+1}\mathbf{u}(x)\, dz\, dx.$$

Replacing $|\delta_z \mathbf{u}(x)| \leq |z| \|\mathbf{u}\|_\infty$, the rest of the term is estimated exactly as $J_{2,2}$.

To summarize, we have obtained the bound:

$$\int_{\mathbb{T}^n} C_\phi(\mathbf{u}, \partial^{m+1}\rho) \cdot \partial^{m+1}\mathbf{u}\, dx \leq \varepsilon \rho_- \|\mathbf{u}\|_{\dot{H}^{m+1+\alpha/2}}^2 + \rho_-^{-1} \|\nabla \mathbf{u}\|_\infty^2 Y_m.$$

Let us now examine the rest of the commutators $C_\phi(\partial^l \mathbf{u}, \partial^{m+1-l}\rho)$ for $l = 1, \dots, m$. After symmetrization we obtain

$$\int_{\mathbb{T}^n} C_\phi(\partial^l \mathbf{u}, \partial^{m+1-l}\rho) \cdot \partial^{m+1}\mathbf{u}\, dx$$

$$= \frac{1}{2} \int_{\mathbb{T}^{2n}} \delta_z \partial^l \mathbf{u}(x) \delta_z \partial^{m+1-l}\rho(x) \partial^{m+1}\mathbf{u}(x) \phi(z)\, dz\, dx$$

$$+ \frac{1}{2} \int_{\mathbb{T}^{2n}} \delta_z \partial^l \mathbf{u}(x) \partial^{m+1-l}\rho(x) \delta_z \partial^{m+1}\mathbf{u}(x) \phi(z)\, dz\, dx := J_1 + J_2.$$

The estimates on the new terms, J_1, J_2, are a slightly more sophisticated as we seek to optimize the distribution of L^p-norms inside their components. Notice that the case $l = 1$ corresponds to the term $J_{2,1}$ that appeared previously.

So, let us assume that $l = 1, \ldots, m$. We will distribute the parameters in J_1 as follows:

$$J_1 = \int_{\mathbb{T}^n \times \mathbb{R}^n} \frac{\delta_z \partial^l \mathbf{u}(x)}{|z|^{\frac{n}{p}+\frac{\alpha}{2}+2\delta}} \frac{\delta_z \partial^{m+1-l} \rho(x)}{|z|^{\frac{n}{q}+\frac{\alpha}{2}}} \frac{\partial^{m+1} \mathbf{u}(x)}{|z|^{\frac{n}{2}-\delta}} \frac{1}{|z|^{\frac{n}{r}-\delta}} \, dz \, dx,$$

where $\delta > 0$ is a small parameter to be determined later and $(2, p, q, r)$ is a Hölder quadruple defined by

$$p = 2 \frac{m + \frac{\alpha}{2}}{l - 1 + \frac{\alpha}{2}}, \quad q = 2 \frac{m + \alpha - 1}{m - l + \frac{\alpha}{2}}, \quad \frac{1}{r} = 1 - \frac{1}{2} - \frac{1}{p} - \frac{1}{q}.$$

The existence of a finite r is warranted by the strict inequality which is verified directly:

$$\frac{1}{2} + \frac{1}{p} + \frac{1}{q} < 1.$$

By the Hölder inequality

$$J_1 \leq \|\mathbf{u}\|_{\dot{W}^{l+\frac{\alpha}{2}+2\delta, p}} \|\rho\|_{\dot{W}^{m+1-l+\frac{\alpha}{2}, q}} \|\mathbf{u}\|_{\dot{H}^{m+1}}.$$

Let us apply the following Gagliardo-Nirenberg inequalities to all the terms:

$$\|\mathbf{u}\|_{\dot{H}^{m+1}} \leq \|\mathbf{u}\|_{\dot{H}^{m+1+\frac{\alpha}{2}}}^{\frac{2m}{2m+\alpha}} \|\nabla \mathbf{u}\|_2^{\frac{\alpha}{2m+\alpha}} \leq \|\mathbf{u}\|_{\dot{H}^{m+1+\frac{\alpha}{2}}}^{\frac{2m}{2m+\alpha}} \|\nabla \mathbf{u}\|_\infty^{\frac{\alpha}{2m+\alpha}},$$

$$\|\mathbf{u}\|_{\dot{W}^{l+\frac{\alpha}{2}+2\delta, p}} \leq \|\mathbf{u}\|_{\dot{H}^{m+1+\frac{\alpha}{2}}}^{\theta_1} \|\nabla \mathbf{u}\|_\infty^{1-\theta_1},$$

$$\|\rho\|_{\dot{W}^{m+1-l+\frac{\alpha}{2}, q}} \leq \|\rho\|_{\dot{H}^{m+\alpha}}^{\theta_2} \|\nabla \rho\|_\infty^{1-\theta_2},$$

where

$$\theta_1 = \frac{l - 1 + \frac{\alpha}{2} - \frac{n}{p} + 2\delta}{m + \frac{\alpha}{2} - \frac{n}{2}}, \quad \theta_2 = \frac{m - l + \frac{\alpha}{2} - \frac{n}{q}}{m + \alpha - 1 - \frac{n}{2}}.$$

The exponents satisfy the necessary requirements

$$1 \geq \theta_1 \geq \frac{l - 1 + \frac{\alpha}{2} + 2\delta}{m + \frac{\alpha}{2}}, \quad 1 \geq \theta_2 = \frac{m - l + \frac{\alpha}{2}}{m + \alpha - 1},$$

and in fact

$$\theta_1 = \frac{l - 1 + \frac{\alpha}{2}}{m + \frac{\alpha}{2}} + O(\delta).$$

Now, we have

$$J_1 \leq \|\mathbf{u}\|_{\dot{H}^{m+1+\frac{\alpha}{2}}}^{\frac{2m}{2m+\alpha}+\theta_1} \|\rho\|_{\dot{H}^{m+\alpha}}^{\theta_2} \|\nabla\mathbf{u}\|_{\infty}^{\frac{\alpha}{2m+\alpha}+1-\theta_1} \|\nabla\rho\|_{\infty}^{1-\theta_2}.$$

By the generalized Young inequality

$$J_1 \leq \varepsilon\rho_- \|\mathbf{u}\|_{\dot{H}^{m+1+\alpha/2}}^2 + \rho_-^{-1} \|\rho\|_{\dot{H}^{m+\alpha}}^{\theta_2 Q} (\|\nabla\mathbf{u}\|_{\infty}^{\frac{\alpha}{2m+\alpha}+1-\theta_1} \|\nabla\rho\|_{\infty}^{1-\theta_2})^Q.$$

where Q is the conjugate to $\frac{2m}{2m+\alpha}+\theta_1$. We have $\theta_2 Q < 2$ as long as

$$\theta_1 + \theta_2 < 2 - \frac{2m}{2m+\alpha}. \tag{7.31}$$

We in fact have an even stronger inequality, $\theta_1 + \theta_2 < 1$, provided δ is small enough. So, we arrived at

$$J_1 \leq \varepsilon\rho_- \|\mathbf{u}\|_{\dot{H}^{m+1+\alpha/2}}^2 + \rho_-^{-1} p_N(\|\nabla\rho\|_\infty, \|\nabla\mathbf{u}\|_\infty) Y_m,$$

for some polynomial p_N.

Finally, moving on to J_2, we distribute the exponents as follows:

$$J_2 \leq \int_{\mathbb{T}^n \times \mathbb{R}^n} \frac{|\delta_z \partial^l u(x)|}{|z|^{\frac{n}{p}+2\delta+\frac{\alpha}{2}}} \frac{|\partial^{m+1-l}\rho(x)|}{|z|^{\frac{n}{q}-\delta}} \frac{|\delta_z \partial^{m+1} u(x)|}{|z|^{\frac{n}{2}+\frac{\alpha}{2}}} \frac{1}{|z|^{\frac{n}{r}-\delta}} \, dz \, dx$$

$$\leq \|u\|_{\dot{W}^{l+\delta+\frac{\alpha}{2},p}} \|\rho\|_{\dot{W}^{m+1-l,q}} \|u\|_{\dot{H}^{m+1+\frac{\alpha}{2}}}.$$

Here, we choose (r, p, q, δ) as follows:

$$q = 2\frac{m+\alpha-1}{m-l}, \quad p = 2\frac{m+\frac{\alpha}{2}}{l-1+\frac{\alpha}{2}}, \quad \frac{1}{r} = 1 - \frac{1}{2} - \frac{1}{p} - \frac{1}{q},$$

and δ is small. With these choices we proceed via the Gagliardo-Nirenberg inequalities:

$$\|\mathbf{u}\|_{\dot{W}^{l+2\delta+\frac{\alpha}{2},p}} \leq \|\mathbf{u}\|_{\dot{H}^{m+1+\frac{\alpha}{2}}}^{\theta_1} \|\nabla\mathbf{u}\|_{\infty}^{1-\theta_1},$$

$$\|\rho\|_{\dot{W}^{m+1-l,q}} \leq \|\rho\|_{\dot{H}^{m+\alpha}}^{\theta_2} \|\nabla\rho\|_{\infty}^{1-\theta_2},$$

where

$$\theta_1 = \frac{l-1+\frac{\alpha}{2}+2\delta}{m+\frac{\alpha}{2}-\frac{n}{2}} = \frac{l-1+\frac{\alpha}{2}}{m+\frac{\alpha}{2}} + O(\delta), \quad \theta_2 = \frac{m-l}{m+\alpha-1}.$$

Now, to achieve the bound

$$J_2 \leq \varepsilon\rho_- \|\mathbf{u}\|^2_{\dot{H}^{m+1+\alpha/2}} + \rho_-^{-1} p_N(\|\nabla\rho\|_\infty, \|\nabla\mathbf{u}\|_\infty) Y_m,$$

we have to make sure that $\theta_1 + \theta_2 \leq 1$, which is true for small δ.

We have proved the following a priori bound on $\mathbf{u}$:

$$\frac{d}{dt}\|\mathbf{u}\|^2_{\dot{H}^{m+1}} \leq -\frac{1}{2}\rho_- \|\mathbf{u}\|^2_{\dot{H}^{m+1+\alpha/2}} + \rho_-^{-1} p_N(\|\nabla\rho\|_\infty, \|\nabla\mathbf{u}\|_\infty) Y_m.$$

Together with the previously established bounds, we obtain

$$\frac{d}{dt}Y_m \leq -\frac{1}{2}\rho_- \|\mathbf{u}\|^2_{\dot{H}^{m+1+\alpha/2}} + \rho_-^{-1} p_N(\|\nabla\rho\|_\infty, \|\nabla\mathbf{u}\|_\infty, \|e\|_\infty) Y_m.$$

This of course implies a Riccati inequality, provided $m > \frac{n}{2} + 1$:

$$\frac{d}{dt}Y_m \leq C Y_m^N,$$

and provides an a priori bound independent of the viscosity coefficient. Thus, we can extend it to an interval independent of ε as well. By a compactness argument similar to the smooth kernel case, we obtain a local solution in the same class as initial data and

$$\mathbf{u} \in L^2([0, T_0); \dot{H}^{m+1+\alpha/2}).$$

In addition, we obtain a continuation criterion—as long as $\|\nabla\rho\|_\infty$, $\|\nabla\mathbf{u}\|_\infty$, $\|e\|_\infty$ remain bounded on $[0, T_0)$, the solutions can be extended beyond T_0. However everything is reduced to a control over the first two quantities, because $\|e\|_\infty$ remains bounded as long as $\|\nabla\mathbf{u}\|_\infty$ is in view of (7.28).

It is clear from the proof that (7.20) can be replaced with an integrability condition with some high power depending on m, n, α.

7.3   Notes and References

Local existence and small initial data results for smooth models appeared in various contexts in works of Ha et al. [47, 48]. Our Theorem 7.1 with continuation criterion in terms of divergence (7.9) was proved in [65], although criteria of the type (7.9) were known in the context of hyperbolic conservation laws; see Grassin [42], Poupaud [85] and references therein.

Equation (7.18) was found in many different contexts, for example, as a 1D model of the surface quasi-geostrophic equation [24], as a model for porous media with nonlocal pressure [11]. The blowup was proved by Castro and Cordoba in [19]. Remarkably, this blowup example can be extended to all $0 < \alpha < 2$ in the context of the Euler alignment system as shown by Arnaiz and Castro [4]. Earlier Tan [100] demonstrated growth of $\|\rho\|_{C^1}$ as $t \to \infty$ for a similar density configuration with 1-point vacuum.

Although in 1D local existence for singular models appeared in the first papers by Shvydkoy and Tadmor [93] for $\alpha \geq 1$ and Do et al. [39] for $0 < \alpha < 1$, it was not until now that the multidimensional case was addressed with a proper continuation criterion; see Lear and Shvydkoy [66]. For singular topological models, local existence was presented in Reynolds and Shvydkoy [88], and it is much more technically involved due to active dependencies on the density in the kernel. No practical continuation criterion has been established in this case yet.

The issues of a sharp threshold condition for multi-D smooth models and global well-posedness for non-vacuous periodic solutions for singular models remain outstanding and important open problems. The models have not been studied numerically extensively. However, Mao et al. [71] provided evidence that at least for $\alpha = 0.5$ in 1D and $\alpha = 1.2$ in 2D, the singular models produce a very close match with the discrete dynamics. Those particular values were gathered through machine learning techniques and the use of data related to the StarFlag project [20, 21, 23]. Numerical simulations were performed on a time interval long enough to observe flocking states.

Chapter 8
One-Dimensional Theory

In dimension 1, the most complete regularity theory of alignment models is available. In this chapter we discuss some of its highlights.

In 1D the system is given by

$$\begin{cases} \rho_t + (\rho u)_x = 0, \\ u_t + u u_x = \displaystyle\int_\Omega \phi(x - y)(u(y) - u(x))\rho(y)\,dy \end{cases} \quad (x, t) \in \Omega \times \mathbb{R}_+.$$

$$(8.1)$$

Here Ω is either $\mathbb{R}$ or the periodic domain $\mathbb{T}$. The underlying theme in this chapter will be to relate regularity and flocking behavior of the system to the new conserved quantity which is available in dimension 1 and a few other exceptional multi-D cases:

$$e = u_x + \mathcal{L}_\phi \rho, \qquad (8.2)$$

where $\mathcal{L}_\phi$ takes the form of either of the integral representations (6.3) or (6.4) depending on the context. In disguise this is nothing but the continuous analogue of (the derivative of) the discrete conserved quantities (2.33), and as such it satisfies a continuity law similar to the density equation. The physical role of e will be illuminated later in Sect. 8.2.

8.1 Smooth Kernels: Critical Thresholds and Stability

The system (8.1) possesses an extra conservation law provided ϕ is a convolution kernel, $\phi = \phi(x - y)$. To see this we note that the alignment term in the velocity equation is given by the commutator:

© Springer Nature Switzerland AG 2021
R. Shvydkoy, *Dynamics and Analysis of Alignment Models of Collective Behavior*,
Nečas Center Series, https://doi.org/10.1007/978-3-030-68147-0_8

$$\mathcal{C}_\phi(u, \rho) = \phi * (u\rho) - u\,\phi * \rho.$$

Using this commutator structure and by elementary manipulation with the equations, we obtain

$$e_t + (ue)_x = 0, \quad e = u_x + \phi * \rho. \tag{8.3}$$

Note that the Burgers' transport tends to create shocks, while the alignment force creates a smoothing counterbalance mechanism. One would expect then that a threshold condition would separate singular behavior from regular. For the pure Burgers equation, such a condition is provided by a positive initial slope $u_x \geq 0$. In view of the regularizing additional effect of the alignment, such a condition can be relaxed to $e_0 \geq 0$. To see this let us rewrite the e-equation as a nonautonomous logistic ODE along characteristics:

$$\dot{e} = e(\phi * \rho - e). \tag{8.4}$$

It is clear that the sign of e will be preserved pointwise. So, if $e_0(x_0) < 0$ at some point x_0, then $\dot{e} < -e^2$, and consequently the solution blows up in finite time. On the other hand, supposing $e_0 \geq 0$, we have $e(t) \geq 0$ for all time, and e remains a priori bounded since $\phi * \rho \leq CM$. This in particular implies that $u_x \in L_{t,x}^\infty$. In view of Theorem 7.1, this ensures global existence of solutions. Moreover, we can also obtain a hydrodynamic version of strong flocking. Let us make this precise.

Theorem 8.1 (Threshold for Global Existence) *Consider the system* (8.1) *on* $\mathbb{R}$ *or* $\mathbb{T}$ *with smooth kernel. For any initial condition* $(u_0, \rho_0) \in H^m \times (L_+^1 \cap W^{1,\infty})$, $m > \frac{n}{2} + 2$, *which satisfies the threshold condition* $e_0 \geq 0$, *there exists a unique global solution* $(u, \rho) \in C_w([0, \infty); H^m \times (L_+^1 \cap W^{1,\infty}))$. *If* $e_0(x) < 0$ *at some point, then the solution blows up in finite time.*

Furthermore, suppose that ϕ *has a heavy tail. Let the initial flock have compact support* Supp ρ_0 *and* $e_0 > 0$ *on the support of the flock. Then the solution flocks strongly in the following sense: there exist* $C, \delta > 0$ *depending on* ϕ, *and initial data, such that the velocity satisfies*

$$\sup_{x \in \mathrm{Supp}\,\rho(t)} |u(x, t) - \bar{u}| + |u_x(x, t)| + |u_{xx}(x, t)| \leq Ce^{-\delta t}, \tag{8.5}$$

and the density ρ *converges to a traveling wave* $\bar{\rho}$ *in the metric of* C^γ *for any* $0 < \gamma < 1$:

$$\|\rho(\cdot, t) - \bar{\rho}(\cdot - \bar{u}t)\|_{C^\gamma} \leq Ce^{-\delta t}, \quad t > 0. \tag{8.6}$$

Proof The global existence has already been proved in the preceding remarks.

Let us now assume that we have a global solution with $e_0 > 0$ and ϕ has a heavy tail. We know from Theorem 6.1 that the diameter of the flock will remain finite, $\overline{\mathcal{D}}$.

Then we can estimate the convolution from below: for any $x \in \operatorname{Supp} \rho(t)$:

$$\phi * \rho(t, x) \geq \phi(\overline{\mathcal{D}})M := c_0.$$

Solving the logistic ODI: $\dot{e} \geq e(c_0 - e)$, we find that starting from some time t_0 for all $t > t_0$ and $x \in \operatorname{Supp} \rho(t)$, we have $e(x, t) \geq c_0/2$. With this in mind, let us write the equation for u_x as follows:

$$(\partial_t + u\partial_x)u_x = \int_{\mathbb{R}} \phi'(x - y)(u(y) - u(x))\rho(y)\, dy - e(x)u_x(x). \tag{8.7}$$

We already know from Theorem 6.1 that the velocity fluctuations are decaying with exponential rate. Hence, the integral above will be bounded by $\|\phi'\|_\infty M E(t)$, where we denote by E any quantity that shows exponential decay. Thus, multiplying (8.7) with u_x and evaluating at the maximum over $\operatorname{Supp} \rho(t)$ (note that in Lagrangian coordinate this corresponds to evaluating at the maximum over the fixed domain $\operatorname{Supp} \rho_0$; hence Rademacher's Lemma 2.2 applies), we obtain

$$\frac{d}{dt}\|u_x\|_{L^\infty(\operatorname{Supp} \rho(t))} = E(t) - \frac{1}{2}c_0\|u_x\|_{L^\infty(\operatorname{Supp} \rho(t))}.$$

This implies the desired result by integration. In view of (7.7), the density enjoys a pointwise global bound:

$$\sup_{t>0} \|\rho(\cdot, t)\|_\infty = R < \infty. \tag{8.8}$$

For the second derivative, we have

$$(\partial_t + u\partial_x)u_{xx} + 2u_x u_{xx}$$

$$= \int_{\mathbb{R}} \phi''(x - y)(u(y) - u(x))\rho(y)\, dy - 2u_x\phi' * \rho - eu_{xx}.$$

We see that the integral term as well as u_x is of type $E(t)$ in view of the previously established bounds. Note also that $|\phi_x * \rho| \leq \|\phi_x\|_\infty M$. So, we obtain

$$\frac{d}{dt}\|u_{xx}\|_{L^\infty(\operatorname{Supp} \rho(t))} = E(t) - \frac{1}{2}c_0\|u_{xx}\|_{L^\infty(\operatorname{Supp} \rho(t))}. \tag{8.9}$$

This implies exponential decay. Moving to the density, we have

$$(\partial_t + u\partial_x)\rho_x = -2u_x\rho_x - u_{xx}\rho = E\rho_x + E. \tag{8.10}$$

This shows that ρ_x remains uniformly bounded. Now, to establish strong flocking, we pass to the moving frame $x - \bar{u}t$ and write the continuity equation in new coordinates:

$$(\partial_t + \bar{u}\partial_x)\rho = -(u - \bar{u})\rho_x - u_x\rho = E. \tag{8.11}$$

This shows that $\rho(t)$ is Cauchy in t in the metric of L^∞. Hence, there exists $\bar{\rho} \in L^\infty$ such that $\|\rho(t) - \bar{\rho}\|_\infty = E(t)$. Since ρ' is uniformly bounded, this also shows that $\bar{\rho}$ is Lipschitz. Convergence in C^γ, $\gamma < 1$, follows by interpolation.   □

Using the stability estimate (6.19), we can easily conclude stability of the limiting flock distributions in the sense of the KR metric. Indeed, if $\rho \to \bar{\rho}$ in C^γ, then certainly $W_1(\rho, \bar{\rho}) \to 0$. So, if we start from two flocks with the same mass and momentum and bounded supports, then using (6.19) we obtain (by translation invariance of the KR metric)

$$W_1(\bar{\rho}', \bar{\rho}'') \leq C[W_1(\rho_0', \rho_0'') + \|\mathbf{u}_0' - \mathbf{u}_0''\|_{L^\infty(\Omega)}], \tag{8.12}$$

where $\Omega = \text{Supp}\, \rho_0' \cup \text{Supp}\, \rho_0''$. This establishes a direct stability control of limiting flocks with respect to the initial perturbation. To bootstrap the stability estimate to a higher regularity class, let us note that in the course of the proof of Theorem 8.1, we established a global bound on $\|\rho_x\|_\infty$ by a constant depending on the initial data. Thus, $\bar{\rho}$ will remain in $W^{1,\infty}$ with a similar bound. Interpolating between $W^{-1,\infty}$ (which is equivalent to the KR metric, see Remark 6.1) and $W^{1,\infty}$ gives a bound in the Hölder class C^γ, $\gamma < 1$:

$$\|\bar{\rho}' - \bar{\rho}''\|_{C^\gamma} \lesssim [W_1(\rho_0', \rho_0'') + \|\mathbf{u}_0' - \mathbf{u}_0''\|_{L^\infty(\Omega)}]^{\frac{1-\gamma}{2}}. \tag{8.13}$$

8.2 Entropy. Csiszár-Kullback Inequality. Distribution of the Limiting Flock

Although in the case of symmetric communication the limiting velocity is determined from the initial condition by use of momentum conservation, the limiting shape of the density profile $\bar{\rho}$ is an emergent quantity which is not known a priori. Yet it is easy to notice that the e-quantity is somehow involved in determining $\bar{\rho}$. Let us assume in this section that $e = u_x + \mathcal{L}_\phi\rho$, where $\mathcal{L}_\phi$ is defined by (6.4) in either the smooth or singular case. Let us write the continuity equation with the use of e:

$$\rho_t + u\rho_x + e\rho = \rho\mathcal{L}_\phi\rho. \tag{8.14}$$

Let us assume for a moment that ϕ is a smooth absolute kernel on $\mathbb{T}$, i.e., $\inf_{\mathbb{T}} \phi = c_0 > 0$. Suppose that $e_0 = 0$ and hence $e(t) = 0$ for all time. Note that in this case, $u_x + \phi * \rho \geq \rho(x)\|\phi\|_1 \geq 0$, so the global solution exists. Then

$$\rho_t + u\rho_x = \rho \mathcal{L}_\phi \rho,$$

and consequently, ρ obeys the maximum principle. Let us assume that there is no vacuum $\rho_-(0) > 0$. Let us then write the equation for the new quantity $\ln \rho$:

$$(\ln \rho)_t + u(\ln \rho)_x = \mathcal{L}_\phi \rho.$$

Evaluating at a point of minimum, we obtain

$$(\ln \rho_-)_t = \int_{\mathbb{T}} \phi(x, y)(\rho(y) - \rho_-)\, dy \geq c_0(M - 2\pi\rho_-),$$

and at the maximum

$$(\ln \rho_+)_t = \int_{\mathbb{T}} \phi(x, y)(\rho(y) - \rho_+)\, dy \leq c_0(M - 2\pi\rho_+).$$

Subtracting the two we obtain

$$\frac{d}{dt} \ln \frac{\rho_+}{\rho_-} \leq -2\pi c_0(\rho_+ - \rho_-) \leq -2\pi c_0 \rho_-(0)\left(\frac{\rho_+}{\rho_-} - 1\right) \leq -2\pi c_0 \rho_-(0) \ln \frac{\rho_+}{\rho_-}.$$

We conclude that

$$\ln \frac{\rho_+}{\rho_-} \leq c_1 e^{-c_2 t}.$$

Since the maximum also stays bounded, we have inequality

$$\ln \frac{\rho_+}{\rho_-} \geq c\left(\frac{\rho_+}{\rho_-} - 1\right).$$

So, we conclude that the density flattens out exponentially fast to a uniformly distributed state $\bar{\rho} = \frac{1}{2\pi}M$.

In view of this computation, we can see that e is directly responsible for the flattening of the density. It turns out that, first, a similar result is true for local kernels and even vacuous solutions. And second, in general the size of e per mass, i.e., the quotient $q = \frac{e}{\rho}$, measures how far $\bar{\rho}$ is from the uniform distribution. Thus, e plays the role of a topological entropy of the flock—a measure of disorder. We will address this interpretation in the next theorem.

Theorem 8.2 *Let (ρ, u) be a smooth solution to the system (8.1) on the 1D torus $\mathbb{T}$ and ϕ is a smooth local kernel:*

$$\phi(r) \geq \lambda \mathbb{1}_{r < r_0}.$$

If $e_0 = 0$, then

$$\|\rho(t) - \bar{\rho}\|_1 \le c_1(\|\rho_0\|_2)e^{-c_2(\lambda, r_0, M, \|\rho_0\|_\infty)t}, \tag{8.15}$$

where $\bar{\rho} = \frac{1}{2\pi}M$.
  In general, provided $\|q_0\|_\infty < \|\phi\|_1$, one has

$$\limsup_{t \to \infty} \|\rho(\cdot, t) - \bar{\rho}\|_1 \le \frac{M\|q_0\|_\infty\|\phi\|_\infty}{\lambda c(r_0)(\|\phi\|_1 - \|q_0\|_\infty)}. \tag{8.16}$$

Let us note that the dependence on $\|q_0\|_\infty$ is linear for small values. At the same time, the bound is inversely proportional to the strength λ, which shows the stabilizing effect of communication on the structure of the flock. Let us note that q satisfies the transport equation:

$$\partial_t q + u q_x = 0, \tag{8.17}$$

and hence, the value of $\|q\|_\infty$ is preserved for all time.
  The proof will be split in several steps. First, let us recall a very useful tool.

Lemma 8.1 (The Csiszár-Kullback Inequality) *Let us consider two functions $f \ge 0$, $g > 0$ on a measure space (Ω, Σ, μ), such that*

$$\int_\Omega f(x)\,d\mu(x) = \int_\Omega g(x)\,d\mu(x).$$

Then

$$\int_\Omega |f - g|^2 \frac{d\mu}{g} \ge \int_\Omega f \log \frac{f}{g}\,d\mu \ge \frac{1}{8}\|f - g\|_1^2. \tag{8.18}$$

Proof Let us start from an elementary inequality:

$$x(x - 1) \ge x \log x \ge (x - 1) + \frac{1}{2}(x - 1)^2 \mathbb{1}_{\{x<1\}}.$$

The upper inequality is elementary:

$$\int_\Omega f \log \frac{f}{g}\,d\mu \le \int_\Omega f\left[\frac{f}{g} - 1\right]d\mu = \int_\Omega (f-g)\left[\frac{f}{g} - 1\right]d\mu = \int_\Omega |f-g|^2 \frac{d\mu}{g}.$$

Let us prove the lower inequality. On the one hand

$$\int_\Omega f \log \frac{f}{g}\,d\mu(x) \ge \underbrace{\int_\Omega (f - g)\,d\mu(x)}_{=0} + \frac{1}{2}\int_{f<g} g(f/g - 1)^2\,d\mu(x),$$

and on the other hand

$$\|f - g\|_1 = \int_{f<g} (g - f)\, d\mu + \int_{g \leq f} (f - g)\, d\mu$$

$$= \int_{f<g} (g - f)\, d\mu - \int_{f<g} (f - g)\, d\mu$$

$$= 2 \int_{f<g} (g - f)\, d\mu.$$

Considering $g\, d\mu$ as a probability measure, we use the Hölder inequality:

$$\int_{f<g} (g - f)\, d\mu = \int_{f<g} (1 - f/g)g\, d\mu \leq \left(\int_{f<g} |1 - f/g|^2 g\, d\mu \right)^{\frac{1}{2}}.$$

Connecting the two assertions produces lower inequality in (8.18).   □

The object of our study will by the relative entropy defined by

$$\mathcal{H} = \int_{\mathbb{T}} \rho \log \frac{\rho}{\bar{\rho}}\, dx = \int_{\mathbb{T}} \rho \log \rho\, dx - M \log \bar{\rho}. \tag{8.19}$$

By the Csiszár-Kullback inequality applied to $f = \rho/M$, $g = 1/2\pi$ on $\mathbb{T}$, we obtain

$$\frac{1}{16\pi} \|\rho - \bar{\rho}\|_{L^1}^2 \leq \bar{\rho}\mathcal{H} \leq \|\rho - \bar{\rho}\|_{L^2}^2. \tag{8.20}$$

The next technical step in the proof of Theorem 8.2 is the equation for the entropy (8.19) which one obtains by testing the continuity equation with $\log \rho + 1$:

$$(\rho \log \rho)_t = \rho_t (\log \rho + 1) = -(\rho u)'(\log \rho + 1)$$

$$= -\rho'(\log \rho + 1)u - \rho u'(\log \rho + 1)$$

$$= -(\rho \log \rho)'u - (\rho \log \rho)u' - \rho u'$$

$$= -[u(\rho \log \rho)]' - \rho u' = -[u(\rho \log \rho)]' - \rho^2 q + \rho \mathcal{L}_\psi \rho.$$

Therefore

$$\frac{d\mathcal{H}}{dt} = \frac{d}{dt} \int_{\mathbb{T}} \rho \log \rho\, dx = - \int_{\mathbb{T}} \rho^2 q\, dx - \int_{\mathbb{T}^2} \phi(x - y)(\rho(x) - \rho(y))\rho(x)\, dx\, dy.$$

Noting that $\int_{\mathbb{T}} \rho q\, dx = \int_{\mathbb{T}} e\, dx = 0$, we can subtract $\bar{\rho}$ from one density in the first integral on the left-hand side. After additionally symmetrizing the last integral, we obtain

$$\frac{d\mathcal{H}}{dt} = - \int_{\mathbb{T}} (\rho - \bar{\rho})\, \rho q\, dx - \frac{1}{2} \int_{\mathbb{T}^2} \phi(x - y)|\rho(x) - \rho(y)|^2\, dx\, dy. \tag{8.21}$$

Next we obtain bounds on the dissipation (enstrophy) term provided by the continuity equation. If our kernel were absolute, it would be easy to get a positive lower bound:

$$\int_{\mathbb{T}^2} \phi(x-y)|\rho(x)-\rho(y)|^2\,dx\,dy \geq (\inf\phi)\int_{\mathbb{T}^2}|\rho(x)-\rho(y)|^2\,dx\,dy \tag{8.22}$$

$$= 2(\inf\phi)\|\rho-\bar{\rho}\|_2^2.$$

Since we have a nontrivial lower bound on the kernel only near the diagonal $\{(x,y)\in\mathbb{T}^2 : |x-y| < r_0\}$, we need a substitute for (8.22) stated in the following lemma.

Lemma 8.2 *The following inequality holds:*

$$\iint_{|x-y|<r_0}|\rho(x)-\rho(y)|^2\,dy\,dx \geq c(r_0)\|\rho-\bar{\rho}\|_2^2. \tag{8.23}$$

Proof Denote by χ any nonnegative bump function supported on $B_{r_0}(0)$, constant on $B_{r_0/2}$ and such that $\int \chi(r)\,dr = 1$. Then on the Fourier side, $\hat{\chi}(0) = 1$ and $|\hat{\chi}(k)| < 1$ for all $k \in \mathbb{Z}\setminus\{0\}$. On the other hand, by the Riemann-Lebesgue Lemma, $\hat{\chi}(k) \to 0$ as $k \to \infty$. Therefore, $|\hat{\chi}(k)| \leq 1 - \varepsilon$ for some $\varepsilon > 0$ depending only on r_0 ($k \neq 0$). Define $\bar{\rho}_{r_0}(x) = \chi * \rho(x)$, so that

$$(\rho-\bar{\rho}_{r_0})\widehat{}(k) = (1-\hat{\chi}(k))\hat{\rho}(k).$$

Hence

$$|(\rho-\bar{\rho}_{r_0})\widehat{}(k)| \geq \varepsilon|\hat{\rho}(k)|, \quad k \in \mathbb{Z}, \ k \neq 0,$$

and $\hat{\rho}(0) = \hat{\bar{\rho}}_{r_0}(0)$. Consequently

$$\|\rho-\bar{\rho}\|_2^2 = \sum_{k\in\mathbb{Z}\setminus\{0\}}|\hat{\rho}(k)|^2 \leq \varepsilon^{-2}\sum_{k\in\mathbb{Z}}|(\rho-\bar{\rho}_{r_0})\widehat{}(k)|^2 = \varepsilon^{-2}\|\rho-\bar{\rho}_{r_0}\|_2^2.$$

By $\int_{\mathbb{T}}\chi\,dx = 1$ and the Minkowski inequality

$$\|\rho-\bar{\rho}_{r_0}\|_2^2 = \left\|\int_{\mathbb{T}}\chi(y)(\rho(\cdot)-\rho(\cdot-y))\,dy\right\|_2^2 \leq \int_{|y|<r_0}\|\rho(\cdot)-\rho(\cdot-y)\|_2^2\,dy$$

$$= \int_{\mathbb{T}}\int_{|z|<r_0}|\rho(x)-\rho(x+z)|^2\,dz\,dx.$$

Combining the above we obtain

$$\|\rho - \bar{\rho}\|_2^2 \leq \varepsilon^{-2} \int_{\mathbb{T}} \int_{|z|<r_0} |\rho(x) - \rho(x+z)|^2 \, dz \, dx.$$

Choosing $c(r_0) = \varepsilon^2$ concludes the proof. □

By virtue of the lemma, the dissipation term has the following lower bound

$$\frac{1}{2} \int_{\mathbb{T}^2} \phi(x-y)|\rho(x) - \rho(y)|^2 \, dx \, dy \geq c\|\rho - \bar{\rho}\|_2^2. \tag{8.24}$$

Now, let us go back and revisit the entropy equation (8.21). We have

$$\frac{d\mathcal{H}}{dt} \leq \|\rho(t)\|_\infty \|q_0\|_\infty \|\rho(\cdot, t) - \bar{\rho}\|_1 - c\|\rho(\cdot, t) - \bar{\rho}\|_2^2$$

$$\leq \|\rho(t)\|_\infty \|q_0\|_\infty \sqrt{16\pi \bar{\rho}\mathcal{H}(t)} - c\bar{\rho}\mathcal{H}(t).$$

Setting $Y = \sqrt{\mathcal{H}}$, we obtain

$$\frac{dY}{dt} \leq \|\rho(t)\|_\infty \|q_0\|_\infty \sqrt{\pi\bar{\rho}} - c\bar{\rho}Y(t).$$

By Grönwall's lemma we arrive at

$$Y(t) \leq Y_0 e^{-c\bar{\rho}t} + \sqrt{\pi\bar{\rho}}\|q_0\|_\infty \int_0^t \|\rho(s)\|_\infty e^{-c\bar{\rho}(t-s)} \, ds. \tag{8.25}$$

It is now easy to reach the conclusion of Theorem 8.2. If $e_0 \equiv 0$, then the second term in (8.25) drops out completely and (8.20) closes this particular case.

For general e, we have

$$\limsup_{t\to\infty} \|\rho(\cdot, t) - \bar{\rho}\|_1 \leq M\|q_0\|_\infty \limsup_{t\to\infty} \int_0^t \|\rho(s)\|_\infty e^{-c\bar{\rho}(t-s)} \, ds$$

$$\leq \frac{\|q_0\|_\infty}{\lambda c(r_0)} \limsup_{t\to\infty} \|\rho(t)\|_\infty. \tag{8.26}$$

Thus the proof of Theorem 8.2 is reduced to estimating the density amplitude.

Lemma 8.3 *Suppose that* $\|q_0\|_\infty < \|\phi\|_1$. *Then one has*

$$\limsup_{t\to\infty} \|\rho(t)\|_\infty \leq \frac{M\|\phi\|_\infty}{\|\phi\|_1 - \|q_0\|_\infty}.$$

Proof First let us observe that if $\dot{X}(t) \leq AX(t)[B - X(t)]$, where A and B are positive constants and $X(t)$ is a positive function, then

$$X(t) \leq \frac{BX(0)}{X(0) + (B - X(0)) \exp(-ABt)}. \tag{8.27}$$

In particular, $\limsup_{t \to \infty} X(t) \leq B$.

Let $\rho_+(t)$ denote the maximum ρ at time t, and let x_+ be a point where the maximum is achieved. Then since $\|q_0\|_\infty < \|\phi\|_1$, one can get an upper bound on $\|\rho(t)\|_\infty$ by integrating the differential inequality derived below.

$$\frac{\mathrm{d}}{\mathrm{d}t}\rho_+(t) = -\rho_+(t)u'(x_+, t)$$

$$= -\rho_+(t)^2 q(x_+, t) + \rho_+(t) \int_{\mathbb{T}} \phi(x_+ - y)(\rho(y, t) - \rho_+(t))\,\mathrm{d}y$$

$$\leq (\|q_0\|_\infty - \|\phi\|_1)\rho_+(t)^2 + M\|\phi\|_\infty \rho_+(t)$$

$$= (\|\phi\|_1 - \|q_0\|_\infty)\rho_+(t)\left[\frac{M\|\phi\|_\infty}{\|\phi\|_1 - \|q_0\|_\infty} - \rho_+(t)\right].$$

In view of (8.27), we arrive at the statement of the lemma.   □

Plugging into (8.26) we conclude (8.16) and the proof of Theorem 8.2 follows.

8.3   Alignment on $\mathbb{T}$ with Degenerate Kernel

On the periodic interval, there is a mechanism for alignment with local communication even for vacuous solutions. It is clear already on the level of particle dynamics—if two agents have not yet aligned and are past their communication range, they would meet at the opposite end of the circle and reestablish communication. The alignment can be established in this case with an adaptation of the corrector method we introduced in Sect. 2.7.

Theorem 8.3 *For any solution of either the discrete or the hydrodynamic system on $\mathbb{T}$ the following holds:*

(i) *For sub-quadratic communication*

$$\lambda \mathbb{1}_{r < r_0} \leq \phi(r) \leq \frac{\Lambda}{r^2}, \tag{8.28}$$

  one has

$$\mathcal{V}_2(t) \leq C\frac{\ln t}{t}, \tag{8.29}$$

  as $t \to \infty$, where C depends only on the initial condition.

(ii) *If the kernel satisfies the more singular assumption*

Fig. 8.1 Slope kernel

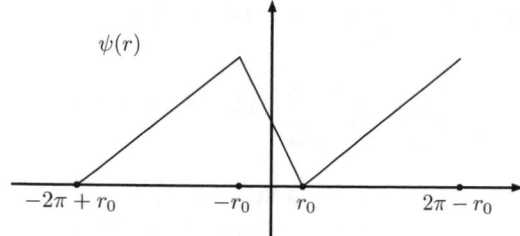

$$\mathbb{1}_{r<r_0}\frac{\lambda}{r^\beta} \le \phi(r) \le \frac{\Lambda}{r^\beta}, \quad \beta > 2, \tag{8.30}$$

then we can only conclude that $\mathcal{V}_2(t) \to 0, \quad t \to \infty.$

Let us note that this result does not improve the rate in the topological models since the variations of the density preclude us from making assumption (8.28).

Proof The proof will be carried out in the discrete case only since the continuous version is entirely similar. We will limit ourselves to providing the necessary modifications in that case.

We go back to the basic energy law (2.40) and construct a corrector functional $\mathcal{G}$ which serves to compensate for the missing interactions. To do that we first define a periodic analogue of the directed distance:

$$d_{ij}(t) = -x_{ij}\,\text{sgn}(v_{ij}) \mod 2\pi,$$

where $x_i, x_j \in [0, 2\pi)$ are viewed on the same coordinate chart. The distance picks up the length of the arch between x_i and x_j which contracts under the evolution of the agents. The distance undergoes jump discontinuities at $x_i = x_j$ and $v_i = v_j$. At any other point, we have

$$\frac{d}{dt}d_{ij} = -|v_{ij}|. \tag{8.31}$$

Next, we define a slope kernel $\psi \ge 0$ as follows (see Fig. 8.1):

$$\psi(x) = \begin{cases} -x + r_0, & -r_0 \le x \le r_0, \\[2mm] \dfrac{r_0}{\pi - r_0}x - \dfrac{r_0^2}{\pi - r_0}, & r_0 < x < 2\pi - r_0, \end{cases}$$

extended periodically on $\mathbb{R}$. Finally, we define the corrector:

$$\mathcal{G}(t) = \frac{1}{N^2}\sum_{i,j=1}^{N} |v_{ij}|\,\psi(d_{ij}).$$

Let us look into the differentiability of $\mathcal{G}$:

$$\frac{\mathrm{d}}{\mathrm{d}t}\mathcal{G} = -\frac{1}{N^2}\sum_{i,j=1}^{N}|v_{ij}|^2\psi'(d_{ij}) + \frac{2}{N^3}\sum_{i,j=1}^{N}\psi(d_{ij})\mathrm{sgn}(v_{ij})\sum_{k=1}^{N}v_{ki}\phi_{ki}.$$

The formula can be justified classically, at those times when there is no jump, i.e., $x_i \neq x_j$ and $v_i \neq v_j$, due to (8.31). When two agents pass each other $x_i = x_j$, we use periodicity of ψ, and when $v_{ij} = 0$, the factor $|v_{ij}|$ vanishes.

We continue

$$\frac{\mathrm{d}}{\mathrm{d}t}\mathcal{G}(t) = \frac{1}{N^2}\sum_{i,j=1}^{N}|v_{ij}|^2\,\mathbb{1}_{|x_{ij}|\leq r_0} - \frac{r_0}{\pi - r_0}\frac{1}{N^2}\sum_{i,j=1}^{N}|v_{ij}|^2\,\mathbb{1}_{|x_{ij}|\geq r_0} + \mathcal{R},$$

where

$$\mathcal{R} = \frac{2}{N^3}\sum_{i,j,k=1}^{N}\psi(d_{ij})\mathrm{sgn}(v_{ij})v_{ki}\phi_{ki}. \qquad (8.32)$$

Symmetrizing over i, k, we obtain

$$\mathcal{R} = \frac{1}{N^3}\sum_{i,j,k=1}^{N}(\psi(d_{ij})\mathrm{sgn}(v_{ij}) - \psi(d_{kj})\mathrm{sgn}(v_{kj}))v_{ki}\phi_{ki}.$$

In the case $v_i \geq v_j \geq v_k$ or $v_i \leq v_j \leq v_k$, the summand is negative, and so we can neglect it. Continuing

$$\mathcal{R} \leq \frac{1}{N^3}\sum_{i,j,k=1}^{N}\left(\psi(d_{ij})\mathrm{sgn}(v_{ij}) - \psi(d_{kj})\mathrm{sgn}(v_{kj})\right)v_{ki}\phi_{ki}\,\mathbb{1}_{v_j>\max(v_i,v_k)}$$

$$+ \frac{1}{N^3}\sum_{i,j,k=1}^{N}\left(\psi(d_{ij})\mathrm{sgn}(v_{ij}) - \psi(d_{kj})\mathrm{sgn}(v_{kj})\right)v_{ki}\phi_{ki}\,\mathbb{1}_{v_j<\min(v_i,v_k)}$$

$$= \frac{1}{N^3}\sum_{i,j,k=1}^{N}\left(\psi(d_{kj}) - \psi(d_{ij})\right)v_{ki}\phi_{ki}\,\mathbb{1}_{v_j>\max(v_i,v_k)}$$

$$+ \frac{1}{N^3}\sum_{i,j,k=1}^{N}\left(\psi(d_{ij}) - \psi(d_{kj})\right)v_{ki}\phi_{ki}\,\mathbb{1}_{v_j<\min(v_i,v_k)}.$$

In the cases $v_j > \max(v_i, v_k)$ and $v_j < \min(v_i, v_k)$, we see that $v_i - v_j$ and $v_k - v_j$ have the same sign so d_{ij} and d_{kj} are computed in the same direction.

So, by the Lipschitz continuity of ψ and by the triangle inequality, we find that $|\psi(d_{ij}) - \psi(d_{kj})| \leq C|x_i - x_k|$. Therefore

$$\mathcal{R} \leq \frac{C}{N^3} \sum_{i,j,k=1}^{N} |x_{ik}| v_{ik} \phi_{ki} = \frac{C}{N^2} \sum_{i,k=1}^{N} |x_{ik}| v_{ik} \phi_{ki}$$

$$\leq \frac{1}{t} \frac{C}{bN^2} \sum_{i,k=1}^{N} |x_{ik}|^2 \phi_{ki} + t \frac{b}{N^2} \sum_{i,k=1}^{N} v_{ik}^2 \phi_{ki}.$$

Let us proceed now under the assumption of (i). Here we obtain

$$\mathcal{R} \leq \frac{c}{t} + bt\mathcal{I}_2.$$

Then the corrector equation becomes

$$\frac{d}{dt}\mathcal{G}(t) \leq a\mathcal{I}_2 + b(t\mathcal{I}_2 - \mathcal{V}_2) + \frac{c}{t}.$$

Let us form another functional:

$$\mathcal{L} = \mathcal{G} + bt\mathcal{V}_2 + a\mathcal{V}_2.$$

It satisfies the inequality $\frac{d}{dt}\mathcal{L} \leq \frac{c}{t}$. Thus, $\mathcal{L}(t) \lesssim \ln t$, and the resulting bound follows.

For part (ii) we use the collision potential (2.36) with the precomputed bound stated in (2.37)

$$\mathcal{R} \leq C \left(\frac{1}{N^2} \sum_{i,k=1}^{N} |x_{ik}|^{2-\beta} |x_{ik}|^{\beta} \phi_{ki} \right)^{1/2} \sqrt{\mathcal{I}_2}$$

$$\lesssim \sqrt{\mathcal{I}_2}\sqrt{C} \leq c_1\sqrt{\mathcal{I}_2(t)} + c_2\sqrt{\mathcal{I}_2(t)} \int_0^t \sqrt{\mathcal{I}_2(s)}\,ds.$$

We can replace by the generalized Young inequality

$$c_1\sqrt{\mathcal{I}_2(t)} \leq \frac{c_3}{t} + bt\mathcal{I}_2(t),$$

and obtain

$$\frac{\mathrm{d}}{\mathrm{d}t}\mathcal{G}(t) \leq a\mathcal{I}_2 + b(t\mathcal{I}_2 - \mathcal{V}_2) + \frac{c_3}{t} + c_2\sqrt{\mathcal{I}_2(t)}\int_0^t \sqrt{\mathcal{I}_2(s)}\,\mathrm{d}s.$$

With $\mathcal{L}$ being defined as before, we continue

$$\frac{\mathrm{d}}{\mathrm{d}t}\mathcal{L} \lesssim \frac{c_3}{t} + c_2\sqrt{\mathcal{I}_2(t)}\int_0^t \sqrt{\mathcal{I}_2(s)}\,\mathrm{d}s.$$

Integrating over $[0, T]$

$$\mathcal{L}(T) \lesssim \mathcal{L}(0) + \ln T + \left(\int_0^T \sqrt{\mathcal{I}_2(s)}\,\mathrm{d}s\right)^2.$$

Thus

$$\mathcal{V}_2(T) \leq \frac{1}{T}\mathcal{L}(T) \lesssim \frac{\ln T}{T} + \frac{1}{T}\left(\int_0^T \sqrt{\mathcal{I}_2(s)}\,\mathrm{d}s\right)^2.$$

The right-hand side tends to zero which can be readily seen by splitting the integral into $(0, T')$ and (T', T), where T' is large.

We already noted that the hydrodynamic version of the result is identical. Let us make some remarks about the proof. We work with the Lagrangian formulation (6.8). As in the discrete case, we define the directed distance:

$$d_{\alpha\beta}(t) = (x(\alpha, t) - x(\beta, t))\,\mathrm{sgn}(v(\beta, t) - v(\alpha, t)) \mod 2\pi,$$

and the corrector with ψ as before

$$\mathcal{G} = \int_{\mathbb{T}^2} |u_{\alpha\beta}|\psi(d_{\alpha\beta})\,\mathrm{d}m_0(\alpha, \beta).$$

We calculate the derivative of $\mathcal{G}$:

$$\frac{\mathrm{d}}{\mathrm{d}t}\mathcal{G} = -\int_{\mathbb{T}^2} |u_{\alpha\beta}|^2\psi'(d_{\alpha\beta})\,\mathrm{d}m_0(\alpha, \beta)$$

$$+ \int_{\mathbb{T}^3} \mathrm{sgn}(u_{\alpha\beta})\psi(d_{\alpha\beta})\phi_{\alpha\gamma}u_{\gamma\alpha}\,\mathrm{d}m_0(\alpha, \beta, \gamma)$$

$$\leq a\mathcal{I}_2 - b\mathcal{V}_2 + \mathcal{R}.$$

Here

$$\mathcal{R} = \int_{\mathbb{T}^3} \mathrm{sgn}(u_{\alpha\beta})\psi(d_{\alpha\beta})\phi_{\alpha\gamma} u_{\gamma\alpha}\, dm_0(\alpha, \beta, \gamma)$$

$$= \frac{1}{2} \int_{\mathbb{T}^3} [\mathrm{sgn}(u_{\alpha\beta})\psi(d_{\alpha\beta}) - \mathrm{sgn}(u_{\gamma\beta})\psi(d_{\gamma\beta})]\phi_{\alpha\gamma} u_{\gamma\alpha}\, dm_0(\alpha, \beta, \gamma)$$

$$\leq \int_{\mathbb{T}^3} (\psi(d_{\alpha\beta}) - \psi(d_{\gamma\beta}))\phi_{\alpha\gamma} u_{\gamma\alpha} \mathbb{1}_{u(\beta) < \min\{u(\alpha), u(\gamma)\}}\, dm_0(\alpha, \beta, \gamma)$$

$$+ \int_{\mathbb{T}^3} (\psi(d_{\gamma\beta}) - \psi(d_{\alpha\beta}))\phi_{\alpha\gamma} u_{\gamma\alpha} \mathbb{1}_{u(\beta) > \max\{u(\alpha), u(\gamma)\}}\, dm_0(\alpha, \beta, \gamma)$$

$$\leq \int_{\mathbb{T}^2} |x_{\alpha\gamma}||u_{\gamma\alpha}|\phi_{\alpha\gamma}\, dm_0(\alpha, \gamma).$$

In case (i) we obtain

$$\mathcal{R} \leq \frac{c}{t} + bt\mathcal{I}_2,$$

and the proof concludes as in the agent-based settings. In case (ii) we consider the collisional potential:

$$\mathcal{C} = \int_{\mathbb{T}^2} \frac{dm_0(\alpha, \beta)}{(|x_{\alpha\beta}| \wedge r_0)^{\beta-2}}.$$

It is well-posed for $\beta < 3$ (in view of also the fact that the density is bounded for regular solutions). A similar computation establishes (2.37), and from this point on the proof proceeds verbatim.   □

In hydrodynamic settings the L^2-based alignment result does not provide sufficient information for pointwise behavior. So, it is desirable to obtain an L^∞-norm-based alignment statement in this context. The mechanism for such an alignment comes from considering regions where the density is non-negligible— here the alignment term works faster than the transport to avoid agent collisions. At the same time if the density is thin, the equation acts as the classical Burgers' equation. So, in order to avoid a blowup, it must have low velocity fluctuations. In other words, it has to be aligned sufficiently well.

Theorem 8.4 *Consider the system* (8.1) *on* $\mathbb{T}$ *with a smooth nontrivial nonnegative kernel. Then any global classical solution aligns:*

$$\|u(t) - \bar{u}\|_\infty \leq C \left(\frac{\ln t}{t}\right)^{\frac{1}{3}}. \tag{8.33}$$

Proof By the Galilean invariance, we can assume throughout that $\bar{u} = 0$. As a consequence of the energy equality and (8.29), we obtain

$$\int_T^\infty \int_{\mathbb{T}^2} \phi_{\alpha\beta} |v(\alpha, t) - v(\beta, t)|^2 \, dm_0(\alpha, \beta) \, dt \le C \frac{\ln T}{T} := \varepsilon.$$

Here we passed to Lagrangian coordinates $v(\alpha, t) = u(x(\alpha, t), t)$. Denote

$$\mathcal{I}_2(\alpha, T) = \int_T^\infty \int_{\mathbb{T}} \phi_{\alpha\beta} |v(\alpha, t) - v(\beta, t)|^2 \, dm_0(\beta) \, dt.$$

So, we have

$$\int_{\mathbb{T}} \mathcal{I}_2(\alpha, T) \, dm_0(\alpha) \le \varepsilon.$$

Let us fix another small parameter $\delta > 0$ and define the "good set":

$$G_\delta(T) = \{\alpha : \mathcal{I}_2(\alpha, T) \le \delta\}.$$

We denote by G_δ^c the complement of G_δ, so that $m_0(G_\delta^c) = M - m_0(G_\delta)$ (recall that M is the total mass of the flock). By the Chebyshev inequality

$$m_0(G_\delta^c) < \frac{\varepsilon}{\delta}. \qquad (8.34)$$

Thus, the good set occupies almost all of the domain provided $\varepsilon \ll \delta$. We now proceed by proving that alignment occurs first on the good set identified above and then on the rest of the torus later in time within a controlled time scale.   □

Lemma 8.4 (Alignment on G_δ) *We have*

$$\sup_{\alpha_1, \alpha_2 \in G_\delta(T), \, t \ge T} |v(\alpha_1, t) - v(\alpha_2, t)| \lesssim \delta^{2/3}.$$

Proof It suffices to establish alignment at time T only because of monotonicity of the $\mathcal{I}_2$-function:

$$\mathcal{I}_2(\alpha, t) \le \mathcal{I}_2(\alpha, T), \quad t > T,$$

which in particular implies that the good sets are increasing in time

$$G_\delta(T) \subset G_\delta(t).$$

Integrating the Euler Alignment system

$$\partial_t v(\alpha, t) = \int_{\mathbb{T}} \phi_{\alpha\beta} v_{\alpha\beta} \, dm_0(\beta)$$

over $[T, t]$ for any $\alpha \in G_\delta$, we obtain

$$|v(\alpha, t) - v(\alpha, T)| \le \int_T^t \int_{\mathbb{T}} \phi_{\alpha\beta} |v_{\alpha\beta}| \, dm_0(\beta) \lesssim \delta\sqrt{t - T}. \tag{8.35}$$

Assume that for some $\alpha_1, \alpha_2 \in G_\delta$, we have

$$v(\alpha_1, T) - v(\alpha_2, T) > U,$$

where U is to be determined later. Then in view of (8.35)

$$v(\alpha_1, t) - v(\alpha_2, t) > \frac{U}{2},$$

so long as

$$t - T \lesssim \frac{U^2}{\delta^2}. \tag{8.36}$$

During this time interval, the corresponding characteristics will undergo a significant displacement:

$$x(\alpha_1, t) - x(\alpha_2, t) \ge x(\alpha_1, T) - x(\alpha_2, T) + \frac{1}{2}U(t - T) \mod 2\pi,$$

where $\frac{1}{2}U(t - T) > 4\pi$ as long as $t - T \gtrsim \frac{1}{U}$. If this is allowed to happen, then the characteristics will find themselves at the separation distance equal to $2\pi = 0$ at some point in time, which means they would collapse. We then obtain

$$\frac{1}{U} \gtrsim \frac{U^2}{\delta^2}, \tag{8.37}$$

which gives $U \lesssim \delta^{2/3}$ as claimed. $\square$

In the next step, we show that our solution aligns at a certain not too remote later time $t > T$.

Lemma 8.5 (Alignment Outside G_δ) *One has, for all $t \gtrsim T + \frac{1}{\delta^{1/3}+(\varepsilon/\delta)^{1/2}}$*

$$\sup_{\alpha\in\mathbb{T}, \gamma\in G_\delta(T)} |v(\alpha, t) - v(\gamma, t)| \lesssim \delta^{1/3} + (\varepsilon/\delta)^{1/2}.$$

Proof Fix $\alpha \in \mathbb{T}$ and $\gamma \in G_\delta(T)$. Let us write

$$\partial_t v(\alpha, t) = \int_{\mathbb{T}} v_{\beta\alpha}\phi_{\alpha\beta} \, dm_0(\beta) = \int_{\mathbb{T}} (v_{\beta\gamma} + v_{\gamma\alpha})\phi_{\alpha\beta} \, dm_0(\beta)$$

$$= (\phi * \rho)(x(\alpha, t), t)v_{\gamma\alpha} + \int_{\mathbb{T}} v_{\beta\gamma}\phi_{\alpha\beta} \, dm_0(\beta).$$

The integral term on the right-hand side above will remain small for all $t \geq T$, by virtue of Lemma 8.4 and (8.34). Indeed

$$\left| \int_{\mathbb{T}} v_{\beta\gamma}(t)\phi_{\alpha\beta} \, dm_0(\beta) \right| = \left| \int_{G_\delta(T)} v_{\beta\gamma}(t)\phi_{\alpha\beta} \, dm_0(\beta) \right|$$

$$+ \left| \int_{G_\delta^c(T)} v_{\beta\gamma}(t)\phi_{\alpha\beta} \, dm_0(\beta) \right|$$

$$\lesssim \delta^{2/3} + \frac{\varepsilon}{\delta}.$$

Thus

$$(\phi * \rho)v_{\gamma\alpha} - \delta^{2/3} - \frac{\varepsilon}{\delta} \leq \frac{d}{dt}v(\alpha, t) \leq (\phi * \rho)v_{\gamma\alpha} + \delta^{2/3} + \frac{\varepsilon}{\delta}. \qquad (8.38)$$

Let us consider a fixed time $t \gtrsim T + \frac{1}{\delta^{1/3}+(\varepsilon/\delta)^{1/2}}$, and assume that $v_{\alpha\gamma}(t) = U > 0$ for some U to be determined later. Let us now reverse the dynamics backwards in time from the moment t. For a time period $[s, t]$, where $T < s < t$, the difference will remain positive $v_{\alpha\gamma}(s) > 0$. On that time period, the right-hand side of (8.38) implies

$$\partial_t v \leq \delta^{2/3} + \frac{\varepsilon}{\delta},$$

and hence

$$v(\alpha, t) - \left(\delta^{2/3} + \frac{\varepsilon}{\delta} \right)(t - s) \leq v(\alpha, s).$$

Simultaneously, by (8.35) applied for $\gamma \in G_\delta$, we obtain

$$|v(\gamma, t) - v(\gamma, s)| \leq \delta(t - s)^{1/2}.$$

In combination with the previous inequality, this implies

$$U - \left(\delta^{2/3} + \frac{\varepsilon}{\delta} \right)(t - s) - \delta(t - s)^{1/2}$$

$$= v_{\alpha\gamma}(t) - \left(\delta^{2/3} + \frac{\varepsilon}{\delta} \right)(t - s) - \delta(t - s)^{1/2}$$

$$\leq v_{\alpha\gamma}(s).$$

We find that

$$v_{\alpha\gamma}(s) \geq \frac{U}{2},$$

as long as $(t - s) \lesssim \frac{U}{\delta^{2/3} + \frac{\varepsilon}{\delta}}$ and $(t - s) \lesssim \frac{U^2}{\delta^2}$. The former condition is more restrictive, unless $U \lesssim \delta^{4/3}$, in which case we have reached our goal. Arguing as in Lemma 8.4, we obtain collision backwards in time, provided $(t - s) \sim 1/U$. This becomes possible if $U \gtrsim \delta^{1/3} + (\varepsilon/\delta)^{1/2}$ on the time interval of length $t - T \gtrsim 1/U$, which is true under the assumption.

Arguing similarly from the opposite end, $v_{\alpha\gamma}(t) = -U < 0$, we obtain the bound from below. $\square$

Lemma 8.5 implies the following quantified global alignment starting from $t \gtrsim T + \frac{1}{\delta^{1/3} + (\varepsilon/\delta)^{1/2}}$:

$$\sup_{\alpha,\gamma \in \mathbb{T}} |v(\alpha, t) - v(\gamma, t)| \lesssim \delta^{1/3} + (\varepsilon/\delta)^{1/2}.$$

Optimization over δ produces the choice $\delta = \varepsilon^{3/5}$. Recalling that $\varepsilon = \frac{\ln T}{T}$, we obtain

$$\sup_{\alpha,\gamma \in \mathbb{T}} |v(\alpha, t) - v(\gamma, t)| \lesssim \left(\frac{\ln T}{T} \right)^{1/5},$$

for $t \sim T + \left(\frac{T}{\ln T} \right)^{1/5} \sim T$. This concludes the proof.

8.4 Singular Models: Global Well-Posedness

In this section we establish global well-posedness of solutions to the Euler alignment system (8.1) on the torus $\mathbb{T}$ for the case of a singular kernel (7.14)–(7.16).

As always in 1D the corresponding entropy will play a key role in establishing regularity of solutions to (8.1):

$$e = u_x + \Lambda_\alpha \rho. \tag{8.39}$$

The main result is the following.

Theorem 8.5 *Suppose $m \geq 3$ and $0 < \alpha < 2$. Let $(u_0, \rho_0) \in H^{m+1}(\mathbb{T}) \times H^{m+\alpha}(\mathbb{T})$, and $\rho_0(x) > 0$ for all $x \in \mathbb{T}$. Then there exists a unique non-vacuous global in time solution to (8.1) in the class:*

$$u \in C_w([0, \infty); H^{m+1}) \cap L^2([0, \infty); \dot{H}^{m+1+\alpha/2}),$$
$$\rho \in C_w([0, \infty); H^{m+\alpha}). \tag{8.40}$$

Moreover, the solution obeys uniform bounds on the density

$$c_0 \leq \rho(x, t) \leq C_0, \quad t \geq 0, \tag{8.41}$$

and flocks to a state $(\bar{u}, \bar{\rho})$, $\bar{\rho} \in H^{m+\alpha}$, *so that*

$$\|u(t) - \bar{u}\|_{W^{2,\infty}} + \|\rho(\cdot, t) - \bar{\rho}(\cdot - \bar{u}t)\|_{C^{\gamma}} \leq Ce^{-\delta t} \qquad t > 0, \ 0 < \gamma < 1. \tag{8.42}$$

Proof According to the well-posedness result stated in Theorem 7.4, we already have a local solution (u, ρ) on a time interval $[0, T_0)$. We proceed in several steps. First, we establish the uniform bounds (8.41) on the density which depend only on the initial conditions. So, such bounds hold uniformly on the available time interval $[0, T_0)$. Next, we invoke results from the theory fractional parabolic equations to conclude that our solution gains Hölder regularity after a short period of time, and the Hölder exponent as well as the bound on the Hölder norm depends on the L^{∞} bound on the solution. Finally, we establish a continuation criterion, much weaker than that of Theorem 7.4—claiming that any Hölder regularity of the density propels higher order norms beyond T_0. Here the case $\alpha = 1$ turns out to be more challenging than the rest of the range.

Let us recall that the ratio $q = e/\rho$ satisfies the transport equation

$$q_t + u q_x = 0, \tag{8.43}$$

which implies the a priori bound

$$\|q(t)\|_{\infty} = \|q_0\|_{\infty} < \infty, \tag{8.44}$$

for any non-vacuous solution.

Step 1: Bounds on the Density   We start by establishing (8.41) on the given time interval.

We can write the continuity equation as follows:

$$\rho_t + u \rho_x = -q \rho^2 + \rho \Lambda_{\alpha}(\rho). \tag{8.45}$$

Let us evaluate at a point x_+ where the maximum of ρ, denoted ρ_+, is reached. We obtain, with the use of (8.44)

$$\frac{\mathrm{d}}{\mathrm{d}t} \rho_+ = -q(x_+, t) \rho_+^2 + \rho_+ \int_{\mathbb{T}} \phi(z)(\rho(x_+ + z, t) - \rho_+) \, \mathrm{d}z$$

$$\leq \|q_0\|_{\infty} \rho_+^2 + \rho_+ \int_{|z| < r} \phi(z)(\rho(x_+ + z, t) - \rho_+) \, \mathrm{d}z$$

$$\leq \|q_0\|_{\infty} \rho_+^2 + \frac{1}{r^{1+\alpha}} \rho_+ (M - 2r\rho_+) = \|q_0\|_{\infty} \rho_+^2 + \frac{1}{r^{1+\alpha}} M \rho_+ - \frac{2}{r^{\alpha}} \rho_+^2.$$

Let us pick r small enough so that $\frac{2}{r^{\alpha}} > \|q_0\|_{\infty} + 1$. Then

$$\frac{d}{dt}\rho_+ \leq -\rho_+^2 + C(M,r)\rho_+,$$ (8.46)

which establishes the upper bound by integration.

As to the lower bound, we argue similarly. Let ρ_- be the minimum value of ρ and x_- a point where such a value is achieved. We have

$$\frac{d}{dt}\rho_- \geq -\|q_0\|_\infty \rho_-^2 + \rho_- \int_{\mathbb{T}} \phi(z)(\rho(x_- + z, t) - \rho_-)\,dz$$ (8.47)

$$\geq -\|q_0\|_\infty \rho_-^2 + \phi_- \rho_-(M - 2\pi\rho_-) = -c_1\rho_-^2 + c_2\rho_-.$$

This readily implies the bound from below. Note that at this point the global communication of the model is crucial: $\phi_- > 0$.

As a consequence of the lower bound on the density, we have a global bound on the entropy:

$$\sup_{t\in[0,T_0)} \|e(t)\|_\infty < \infty.$$ (8.48)

Step 2: Hölder Regularization The representation of the continuity equation in the form (8.45) puts it into the class of forced fractional parabolic equations with bounded drift and force:

$$\partial_t v + u \cdot \nabla v = L[v] + f,$$

where L has kernel

$$K(x, z, t) = \rho(x)\frac{1}{|z|^{1+\alpha}},$$

which is even with respect to z. The bounds on the density provide uniform ellipticity bounds on the kernel $\frac{1}{|z|^{1+\alpha}} \lesssim K(x, z, t) \lesssim \frac{1}{|z|^{1+\alpha}}$.

With these ingredients at hand, the case $\alpha = 1$ falls under the assumptions of Silverstre's results [98] which provides Hölder regularization bound given by

$$\|\rho\|_{C^\gamma(\mathbb{T}\times[T_0/2,T_0))} \leq C(\|\rho\|_{L^\infty(\mathbb{T}\times[0,T_0))} + \|\rho e\|_{L^\infty(\mathbb{T}\times[0,T_0))}),$$ (8.49)

for some $\gamma > 0$.

The case $\alpha < 1$ falls under the same result provided $u \in L^\infty([0, T_0); C^{1-\alpha})$. This is indeed the case as follows from

$$\Lambda_\alpha^{-1}\partial_x u = \Lambda_\alpha^{-1}e - \rho \in L_{t,x}^\infty.$$

Note that $\Lambda_\alpha^{-1}\partial_x$ is a $(1 - \alpha)$-order differential operator.

Finally, for $\alpha > 1$ the Hölder continuity follows from a similar identity for ρ:

$$\Lambda_{\alpha-1}\rho = \Lambda_1^{-1}e - \mathcal{H}u,$$

where $\mathcal{H}$ is the Hilbert transform. Note that it sends functions in L^∞ to $B_{\infty,\infty}^0$. Hence, $\rho \in B_{\infty,\infty}^{\alpha-1} = C^{\alpha-1}$.

Step 3: Continuation and Flocking   The last step is to show that if the density is bounded in C^γ on the time interval $[T_0/2, T_0)$, then the solution remains uniformly in $W^{1,\infty}$, and hence the continuation criterion of Theorem 7.4 applies. While doing so we will keep track of the estimates on the $W^{1,\infty}$-norm with the purpose of obtaining long time asymptotics.

Step 3a: Control over $\rho' = \rho_x$ So, let us start with ρ':

$$\partial_t \rho' + u\rho'' + u'\rho' + e'\rho + e\rho' = \rho'\Lambda_\alpha\rho + \rho\Lambda_\alpha\rho'. \tag{8.50}$$

Using again $u' = e - \Lambda_\alpha\rho$, we rewrite

$$\partial_t \rho' + u\rho'' + e'\rho + 2e\rho' = 2\rho'\Lambda_\alpha\rho + \rho\Lambda_\alpha\rho'.$$

Evaluating at the maximum of $|\rho'|$ and multiplying by ρ', we obtain

$$\frac{d}{dt}|\rho'|^2 + e'\rho\rho' + 2e|\rho'|^2 = 2|\rho'|^2\Lambda_\alpha\rho + \rho\rho'\Lambda_\alpha\rho'. \tag{8.51}$$

Let us note that q' satisfies the continuity equation, and consequently, $\frac{q'}{\rho}$ is transported. So, $|q'| \leq C\rho$ pointwise. For the e-quantity itself, this implies the pointwise bound

$$|e'(x, t)| \leq C(|\rho'(x, t)| + \rho(x, t)). \tag{8.52}$$

Let us note that in order to make pointwise evaluation possible in (8.52), one has to assume the regularity $e' \in C(\mathbb{T})$ which is guaranteed provided $m \geq 2$. With this at hand, and in view of (8.41) and (8.48), we can bound

$$|e'\rho\rho' + 2e|\rho'|^2| \leq C(|\rho'|^2 + |\rho'|).$$

Thus

$$\frac{d}{dt}|\rho'|^2 = C(|\rho'|^2 + |\rho'|) + 2|\rho'|^2\Lambda_\alpha\rho + \rho\rho'\Lambda_\alpha\rho'. \tag{8.53}$$

Due to the bound from below on ρ, we estimate

$$\rho\rho'\Lambda_\alpha\rho' \le c_1 \int_{\mathbb{R}} \frac{(\rho'(x+z) - \rho'(x))\rho'(x+z)}{|z|^{1+\alpha}} \, dz \le -c_2 D_\alpha \rho'(x), \qquad (8.54)$$

where

$$D_\alpha \rho'(x) = \int_{\mathbb{R}} \frac{|\rho'(x) - \rho'(x+z)|^2}{|z|^{1+\alpha}} \, dz.$$

$\square$

Lemma 8.6 (Nonlocal Maximum Principle) *The following pointwise bound holds*

$$D_\alpha \rho'(x) \ge c \frac{|\rho'(x)|^{2+\alpha}}{\|\rho\|_\infty^\alpha}. \qquad (8.55)$$

Proof Fix an $r > 0$ to be determined later. We write

$$D_\alpha \rho'(x) \ge \int_{|z|>r} \frac{|\rho'(x) - \rho'(x+z)|^2}{|z|^{1+\alpha}} \, dz$$

$$\ge \int_{|z|>r} \frac{|\rho'(x)|^2 - 2\rho'(x+z)\rho'(x)}{|z|^{1+\alpha}} \, dz$$

$$= \frac{|\rho'(x)|^2}{r^\alpha} - 2\rho'(x) \int_{|z|>r} \frac{\rho'(x+z)}{|z|^{1+\alpha}} \, dz.$$

Integrating by parts in the last integral, we further estimate

$$D_\alpha \rho'(x) \ge \frac{|\rho'(x)|^2}{r^\alpha} - c_\alpha |\rho'(x)| \|\rho\|_\infty \frac{1}{r^{1+\alpha}}.$$

Choosing $r = C \frac{\|\rho\|_\infty}{|\rho'(x)|}$, where C is large proves the estimate. $\square$

In view of the density bounds we have a priori (8.41), the nonlocal maximum principle yields the following nonlinear bound

$$D_\alpha \rho'(x) \ge c |\rho'(x)|^{2+\alpha}.$$

We arrive at

$$\frac{d}{dt} |\rho'|^2 = C(|\rho'|^2 + |\rho'|) + 2|\rho'|^2 \Lambda_\alpha \rho - c|\rho'|^{2+\alpha} - \frac{1}{2} D_\alpha \rho'(x). \qquad (8.56)$$

The lower order terms $|\rho'|^2 + |\rho'|$ can be absorbed into the dissipation term by the generalized Young inequality:

$$|\rho'|^2 + |\rho'| \le c_\varepsilon + \varepsilon |\rho'|^{2+\alpha},$$

for $\varepsilon > 0$ small. So, it remains to obtain an estimate on the remaining term $|\rho'|^2 \Lambda_\alpha \rho$.

To do that we fix a scale parameter $1 > r > 0$ to be determined later and split the integral representation of the fractional Laplacian into three parts—short-range, midrange, and long-range:

$$\Lambda_\alpha \rho(x) = \int_{|z|<r} [\delta_z \rho(x) - \rho'(x)z] \frac{dz}{|z|^{1+\alpha}} + \int_{r<|z|<1} \delta_z \rho(x) \frac{dz}{|z|^{1+\alpha}}$$

$$+ \int_{|z|>1} \delta_z \rho(x) \frac{dz}{|z|^{1+\alpha}}$$

$$:= I + II + III.$$

For the short-range we use the dissipation directly:

$$|\rho(x+z) - \rho(x) - \rho'(x)z| = \left| \int_0^z (\rho'(x+w) - \rho'(x))\, dw \right| \tag{8.57}$$

$$\leq \sqrt{D_\alpha \rho'(x)} |z|^{1+\frac{\alpha}{2}},$$

so,

$$|I| \leq r^{1-\alpha/2} \sqrt{D_\alpha \rho'(x)}.$$

In the midrange we use the available Hölder continuity (here we can assume without loss of generality that $\gamma < \alpha$):

$$|II| \leq \|\rho\|_{C^\gamma} r^{\gamma-\alpha} \lesssim r^{\gamma-\alpha}.$$

And finally, for the long-range, we simply use the boundedness of ρ:

$$|III| \lesssim \|\rho\|_\infty.$$

The competition occurs only between the short- and midrange terms. Optimizing over r we set $r = (D_\alpha \rho'(x))^{-\frac{1}{2+\alpha+2\gamma}}$ unless this expression is > 1, in which case we have an absolute bound on the dissipation and the proof proceeds trivially. With the established bounds, we obtain the following pointwise estimate:

$$|\Lambda_\alpha \rho(x)| \lesssim c_1 + c_2 (D_\alpha \rho'(x))^{\frac{\alpha-\gamma}{2+\alpha+2\gamma}}.$$

Note that $\frac{\alpha-\gamma}{2+\alpha+2\gamma} < \frac{\alpha}{2+\alpha}$. So, we can use the generalized Young inequality to obtain

$$|\rho'|^2 |\Lambda_\alpha \rho| \lesssim c_\varepsilon + \varepsilon |\rho'|^{2+\alpha} + \varepsilon D_\alpha \rho'(x).$$

Plugging this into (8.56), we arrive at

$$\frac{d}{dt}|\rho'|^2 \leq c_1 - c_2|\rho'|^{2+\alpha}. \tag{8.58}$$

This concludes the proof of uniform bound $\rho \in L^\infty([0, T_0); W^{1,\infty})$.

Step 3b: Conclusion of the Proof in the Case $\alpha < 1$ We start out with an easier case $0 < \alpha < 1$ where control over u' is straightforward from the e-quantity. Indeed, the e-quantity is uniformly bounded by (8.48), while $\Lambda_\alpha \rho \in L^\infty$ simply by $\|\Lambda_\alpha \rho\|_\infty \leq \|\rho'\|_\infty$. So, we obtain a uniform bound on $\|u'\|_\infty$ and hence global existence by Theorem 7.4. However, this argument does not provide a good quantitative estimate on $\|u'\|_\infty$ to conclude flocking. We will seek more precise estimates with the help of a nonlocal maximum principle and further fractional estimates.

Let us write the equation for u', evaluated at the maximum of $|u'|$ and multiplied by u':

$$\frac{d}{dt}|u'|^2 \leq |u'|^3 + u'(x) \int_{\mathbb{R}} \delta_z u'(x) \rho(x+z) \frac{dz}{|z|^{1+\alpha}}$$
$$+ u'(x) \int_{\mathbb{R}} \delta_z u(x) \rho'(x+z) \frac{dz}{|z|^{1+\alpha}}. \tag{8.59}$$

The dissipation term is bounded, as before by

$$u'(x) \int_{\mathbb{R}} \delta_z u'(x) \rho(x+z) \frac{dz}{|z|^{1+\alpha}} \leq -c D_\alpha u'(x).$$

For $D_\alpha u'(x)$ we can derive another two versions of the nonlocal maximum principle similar to (8.55). First, replacing u' with $(u - \bar{u})'$ from the beginning, we obtain the amplitude $\mathcal{A}$ rather than $\|u\|_\infty$ in the denominator:

$$D_\alpha u'(x) \geq c \frac{|u'(x)|^{2+\alpha}}{\mathcal{A}^\alpha(t)}. \tag{8.60}$$

And second, by setting $r = B^{-1/\alpha}$ where B is an arbitrary constant, we obtain

$$D u'(x) \geq c_1 B |u'(x)|^2 - c_2 |u'(x)| \mathcal{A} B^{\frac{1+\alpha}{\alpha}} \geq c_3 B |u'(x)|^2 - c_4 B^{\frac{2+\alpha}{\alpha}} \mathcal{A}^2.$$

So, we obtain the following bound for any $B > 0$:

$$D_\alpha u'(x) \geq B |u'(x)|^2 - c_2 B^{\frac{1+\alpha}{\alpha}} \mathcal{A}^2. \tag{8.61}$$

We now continue with the last term in (8.59). First, let us consider the case $0 < \alpha < 1$. Then

$$u'(x) \int_{\mathbb{R}} \delta_z u(x) \rho'(x+z) \frac{dz}{|z|^{1+\alpha}} \leq u'(x) \int_{|z|<1} \delta_z u(x) \rho'(x+z) \frac{dz}{|z|^{1+\alpha}}$$

$$+ u'(x) \int_{|z|>1} \delta_z u(x) \rho'(x+z) \frac{dz}{|z|^{1+\alpha}}$$

$$\leq c_1 |u'(x)|^2 \|\rho'\|_\infty + c_2 |u'(x)| \mathcal{A}.$$

Picking $B > 2c_1 \|\rho'\|_\infty$ and using (8.61) for half of the dissipation term and (8.60) for the other half, we obtain

$$\frac{d}{dt} |u'|^2 \leq |u'|^3 - c \frac{|u'(x)|^{2+\alpha}}{\mathcal{A}^\alpha(t)} + C|u'(x)|\mathcal{A}(t) + C\mathcal{A}^2(t). \tag{8.62}$$

We already know from the remark in the beginning of this step that $|u'|$ is uniformly bounded in the case $0 < \alpha < 1$. So, we estimate $|u'|^3 \lesssim |u'|^2$ which again gets absorbed in the dissipation term in view of (8.61). Finally

$$|u'(x)|\mathcal{A}(t) \leq |u'(x)|^2 + \mathcal{A}^2(t), \tag{8.63}$$

which again gets absorbed at the cost of adding another $\mathcal{A}^2(t)$. In the end, we arrive at

$$\frac{d}{dt} |u'|^2 \leq -c \frac{|u'(x)|^{2+\alpha}}{\mathcal{A}^\alpha(t)} + C\mathcal{A}^2(t),$$

which gives uniform control over $\|u'\|_\infty$ and implies an exponential rate of convergence to zero as $t \to \infty$. Since in Step 3a we showed that ρ is uniformly bounded in $W^{1,\infty}$, the proof of strong flocking for the density, $\rho \to \bar{\rho}(\cdot - \bar{u}t)$, follows along the lines of Theorem 8.1. Lastly, showing exponential decay of $\|u''\|_\infty$ follows similar estimates on the evolution of the norm $\|u''\|_\infty^2$, and will not be presented here for the sake of brevity. We refer to [94] for full details.

Step 3c: Conclusion of the Proof in Case the $\alpha > 1$ Next, let us consider the case $\alpha > 1$. We absorb the cubic term in (8.59) simply by interpolation:

$$|u'|^3 \leq \varepsilon \frac{|u'(x)|^{2+\alpha}}{\mathcal{A}^\alpha(t)} + c_\varepsilon \mathcal{A}^{\frac{3\alpha}{\alpha-1}}(t).$$

It comes again to estimating the last term in (8.59). In the long range $\{|z| > 1\}$, we estimate it by $|u'(x)|\mathcal{A}(t)$ and treat it as before in (8.63). In the short range, we add and subtract $u'(x)z$. Noting that by (8.57) we obtain

$$u'(x) \int_{|z|<1} [\delta_z u(x) - u'(x)z]\rho'(x+z)\frac{dz}{|z|^{1+\alpha}} \le C|u'(x)|\|\rho'\|_\infty \sqrt{D_\alpha u'(x)}$$

$$\le \varepsilon D_\alpha u'(x) + c_\varepsilon |u'(x)|^2,$$

where the first term is absorbed and the quadratic term is treated as before. For the remaining term we have

$$|u'(x)|^2 \int_{|z|<1} \rho'(x+z)\frac{z\,dz}{|z|^{1+\alpha}} = |u'(x)|^2 \int_{|z|>1} \rho'(x+z)\frac{z\,dz}{|z|^{1+\alpha}}$$

$$+ |u'(x)|^2 \int_{\mathbb{R}} \rho'(x+z)\frac{z\,dz}{|z|^{1+\alpha}}.$$

The integral over $\mathbb{R}$ is nothing other than $\Lambda_\alpha \rho(x)$ which we replace with $e - u'$. We obtain

$$|u'(x)|^2 \int_{|z|<1} \rho'(x+z)\frac{z\,dz}{|z|^{1+\alpha}} \le c_1 |u'(x)|^2 + c_2 |u'(x)|^3.$$

Both terms have been estimated already before. So, we arrive at

$$\frac{d}{dt}|u'|^2 \le -c_1 \frac{|u'(x)|^{2+\alpha}}{A^\alpha(t)} + c_2 A^\beta(t), \quad \beta > 0,$$

and the result follows.

Step 3d: Conclusion of the Proof in the Case $\alpha = 1$ The case $\alpha = 1$ is more involved because dissipation is not sufficient to control the nonlinearity and the two terms in e, u' and $\Lambda_1 \rho$, are in balance. So, in order to proceed, we need to establish an additional uniform estimate on the second derivative of ρ in L^2, $\|\rho''\|_2 \in L^\infty([0, T_0))$. This can be done bypassing any additional information about u.

Assuming that we have proved $\|\rho''\|_2 \in L^\infty([0, T_0))$ with a constant independent of T_0, we can conclude the proof the theorem as follows. First, let us denote $\Lambda = \Lambda_1$. We establish control over $\Lambda\rho$ as follows:

$$\Lambda\rho(x) = \int_{|z|<1} [\delta_z \rho(x) - \rho'(x)z]\frac{dz}{|z|^2} + \int_{1<|z|} \delta_z \rho(x)\frac{dz}{|z|^2}.$$

The second integral is clearly bounded uniformly. Next,

$$|\delta_z \rho(x) - \rho'(x)z| = \left| \int_0^z \int_0^w \rho''(x+y)\,dy \right| \le \|\rho''\|_2 |z|^{3/2}.$$

So, the first integral is bounded by a constant multiple of $\|\rho''\|_2$. This shows that $\Lambda\rho \in L^\infty([0, T_0); L^\infty)$. This is of course natural because by the Sobolev embedding theorem, we even have $H^2 \to W^{3/2,\infty}$.

Having uniform control over $\|\Lambda\rho\|_\infty$, we immediately obtain control over $u' = e - \Lambda\rho$, and hence the global existence follows. But we also note that in this case $|u'|^3 \lesssim |u'|^2$, and hence by interpolation the cubic term in (8.59) can be hidden into dissipation at the cost of adding a power of $\mathcal{A}$, an exponentially decaying quantity. For the last term in (8.59), we do exactly the same computation as in the case $\alpha > 1$, except that for the short range part we estimate

$$|u'(x)|^2 \int_{|z|<1} \rho'(x+z)\frac{dz}{z} = |u'(x)|^2 \int_{|z|<1} [\rho'(x+z) - \rho'(x)]\frac{dz}{z}$$

$$= |u'(x)|^2 \int_{|z|<1} \int_x^{x+z} \rho''(w)\frac{dz}{z}$$

$$\leq c_1|u'(x)|^2\|\rho''\|_2 \leq c_2|u'(x)|^2.$$

This concludes the estimate of all terms in (8.59), and we finish the proof as in the previous two cases.

So, it remains to obtain a uniform estimate on $\|\rho''\|_2$. Let us write the equation for the second derivative of density:

$$\partial_t\rho'' + u\rho''' + u'\rho'' + e''\rho + 3e'\rho' + 2e\rho''$$
$$= 2\rho''\Lambda\rho + 3\rho'\Lambda\rho' + \rho\Lambda\rho''. \tag{8.64}$$

Let us apply the test-function ρ''/ρ. Via routine computation with the use of the density equation, one can observe that

$$\int_{\mathbb{T}} (\partial_t\rho'' + u\rho''' + u'\rho'')\frac{\rho''}{\rho}\, dx = \frac{1}{2}\partial_t \int_{\mathbb{T}} \frac{1}{\rho}|\rho''|^2\, dx.$$

In view of the bounds on the density, we note that $\int \frac{1}{\rho}|\rho''|^2\, dx \sim \|\rho''\|_2^2$. So, it is sufficient to bound the rest of the terms in terms of $\|\rho''\|_2^2$. Considering the last three terms on the left-hand side, let us make one observation: Since q'/ρ is transported, then $(q'/\rho)'$ satisfies the continuity equation, and hence $(q'/\rho)'/\rho$ is transported again. Solving for e'' in this expression results in the pointwise bound:

$$|e''(x, t)| \leq C(|\rho''(x, t)| + |\rho'(x, t)| + \rho(x, t)). \tag{8.65}$$

In order for this bound to make sense, we require $m \geq 3$. With the use of the a priori estimates established so far

$$\int_{\mathbb{T}} (e''\rho + 3e'\rho' + 2e\rho'')\frac{\rho''}{\rho} \, dx \lesssim 1 + \|\rho''\|_2^2.$$

At this point we have (dropping ρ, ρ' that are already bounded)

$$\partial_t \int_{\mathbb{T}} \frac{1}{\rho}|\rho''|^2 \, dx \lesssim 1 + \|\rho''\|_2^2 + \int_{\mathbb{T}} |\rho''|^2|\Lambda\rho| \, dx +$$

$$+ \int_{\mathbb{T}} |\rho''||\Lambda\rho'| \, dx + \int_{\mathbb{T}} \rho''\Lambda\rho'' \, dx \qquad (8.66)$$

$$= 1 + \|\rho''\|_2^2 + I_1 + I_2 + I_3.$$

Clearly, the last term I_3 is dissipative:

$$I_3 \lesssim -\int_{\mathbb{T}} D_\alpha\rho''(x) \, dx - \frac{1}{\|\rho'\|_\infty}\int_{\mathbb{T}} |\rho''|^3 \, dx = -\|\rho''\|_{\dot{H}^{\frac{1}{2}}}^2 - \|\rho''\|_3^3,$$

where in the latter we dropped $\frac{1}{\|\rho'\|_\infty}$ from inside the integral since this term is bounded from below.

To tackle I_1 we fix $\varepsilon > 0$ small and split the fractional Laplacian:

$$\Lambda\rho(x) = \int_{|z|<\varepsilon} [\delta_z\rho(x) - \rho'(x)z]\frac{dz}{|z|^2} + \int_{\varepsilon<|z|} \delta_z\rho(x)\frac{dz}{|z|^2} \le \varepsilon^{1/2}\|\rho''\|_2 + c_\varepsilon.$$

So,

$$I_1 \le \varepsilon^{1/2}\|\rho''\|_2^3 + c_\varepsilon\|\rho''\|_2^2 \le \varepsilon^{1/2}\|\rho''\|_3^3 + c_\varepsilon\|\rho''\|_2^2.$$

The cubic term gets absorbed by dissipation for small ε.

For I_2 we simply use the Hölder inequality:

$$|I_2| \le \|\rho''\|_2\|\Lambda\rho'\|_2 \lesssim \|\rho''\|_2^2.$$

So,

$$\partial_t \int_{\mathbb{T}} \frac{1}{\rho}|\rho''|^2 \, dx \lesssim 1 + \|\rho''\|_2^2 - \|\rho''\|_3^3. \qquad (8.67)$$

This finishes the proof.

A few remarks are in order. First, the global existence part of the theorem holds for local kernels as well. In other words if we only know that

$$\phi(z) \sim \frac{1}{|z|^{1+\alpha}}, \quad \text{for } |z| < r_0,$$

the proof goes through as long as we can establish bounds on the density from above and below. The bound from above follows the steps of the proof for the global kernels. Indeed, we only need the singularity in the short range to be able to pick small parameter r to achieve (8.46). As to the bound from below, unfortunately we loose much more. However, we can still find an algebraic bound from below for any nonnegative kernel. Indeed, arguing as in (8.47) we simply drop the integral term and arrive at

$$\frac{\mathrm{d}}{\mathrm{d}t}\rho_- \geq -c_1\rho_-^2,$$

which implies

$$\rho(x, t) \gtrsim \frac{1}{1+t}, \qquad \forall \phi \geq 0. \tag{8.68}$$

This, of course, is not sufficient to establish exponential alignment in the local case but is sufficient to apply the continuation argument. We do however recover weak flocking as a consequence of Theorem 8.3. Let us record these observations in the following theorem.

Theorem 8.6 *Suppose the kernel is given by*

$$\phi(z) = \frac{h(|z|)}{|z|^{1+\alpha}}, \quad 0 < \alpha < 2.$$

Suppose for some $m \geq 3$, $(u_0, \rho_0) \in H^{m+1}(\mathbb{T}) \times H^{m+\alpha}(\mathbb{T})$ and $\rho_0(x) > 0$ for all $x \in \mathbb{T}$. Then there exists a unique non-vacuous global in time solution to (8.1) in the class

$$u \in C_w([0, \infty); H^{m+1}) \cap L^2_{\mathrm{loc}}([0, \infty); \dot{H}^{m+1+\alpha/2}),$$

$$\rho \in C_w([0, \infty); H^{m+\alpha}).$$

Moreover, the solution obeys the following bounds on the density:

$$\frac{c_0}{1+t} \leq \rho(x, t) \leq C_0, \quad t \geq 0, \tag{8.69}$$

and weak alignment $\mathcal{V}_2(t) \to 0$, as $t \to \infty$.

8.5   Notes and References

Critical threshold conditions for global regularity of Euler alignment models first appeared in Tan and Tadmor's work [99], although the sharp criterion of

Theorem 8.1 was found later in Carrillo et al. [15]. The concept of strong flocking (8.6) was first introduced and proved actually for singular models on $\mathbb{T}$ in [94, 95] and in the open space in the context of smooth multi-scale models in [96]. We have presented the adapted proof for the mono-scale case.

The role of the e-quantity as an entropy and Theorem 8.2 were proved in Leslie and Shvydkoy [70]. The cases of singular and topological models were also included in that study. The corrector method on $\mathbb{T}$ and Theorem 8.3 was proved in [38].

Singular models and the proof of Theorem 8.5 first appeared in Shvydkoy and Tadmor [93] for the case $\alpha \geq 1$ and shortly after in Do, Kiselev, Ryzhik, and Tan for the case $0 < \alpha < 1$. Later the whole range of α together with strong flocking was covered in [94, 95]. The approach presented here was taken from [93]. The work of Do et al. relies on an adaptation of the modulus of continuity method used to make several breakthroughs in the regularity theory of critical fractional diffusion equations such as critical SQG by Kiselev, Nazarov, Volberg [63]. The latter enjoyed several other different proofs using De Giorgi's method by Caffarelli and Vasseur [10] and the nonlocal maximum principle by Constantin and Vicol [28] which we also used in our proof. An interesting extension of Theorem 8.5 as well as Hölder regularization to the case $\alpha = 0$ was obtained in An and Ryzhik [2]. Other extensions include the construction of weak and strong solutions for the singular model with external force by Leslie [68] and global existence of classical solutions with potential attraction/repulsion forces by Kiselev and Tan [62].

The regularity of topological models with singular kernels given by

$$\phi(x, y) = \frac{h(x - y)}{\mathrm{d}^\tau(x, y)|x - y|^{1+\alpha-\tau}}$$

was developed in [97] with the use of a blend of techniques including De Giorgi's method, fractional Schauder estimates, and Silvestre regularization results [98] depending on the range of α. The final result states global existence in Sobolev classes provided $\tau \leq \alpha$ or if $\tau > \alpha$ then under an additional smallness condition. The $\tau \leq \alpha$ condition is needed to prove global a priori bounds on the density, whereby the metric component of the kernel plays a decisive role.

Chapter 9
Global Solutions to Multidimensional Systems

Global existence of the Euler alignment system in dimension 2 and higher is only partially resolved at the moment. In the smooth kernel case, no sharp threshold condition is known, and in the singular case, the question is wide open. However, certain classes of special solutions do permit global existence. Those include small data in the sense of either velocity fluctuations, spectral gap of the symmetric velocity strain tensor, or special classes respecting symmetries of the system.

9.1 Unidirectional Flocks and Their Stability

One class of solutions that behaves like 1D is the class of unidirectionally oriented flocks. These are given by (see Fig. 9.1)

$$\mathbf{u}(x,t) = u(x,t)\,\mathbf{d}, \quad \mathbf{d} \in \mathbb{S}^{n-1}, \quad u : \mathbb{R}^n \times \mathbb{R}^+ \to \mathbb{R}. \tag{9.1}$$

The same conservation law holds for the entropy:

$$e = \mathbf{d} \cdot \nabla u + \phi * \rho, \quad \partial_t e + \nabla \cdot (e\mathbf{u}) = 0,$$

although in this case the entropy does not control the full gradient of the velocity field. Nonetheless, one can develop a theory fully analogous to the 1D theory in the smooth communication case, which is what we will cover in this section.

© Springer Nature Switzerland AG 2021
R. Shvydkoy, *Dynamics and Analysis of Alignment Models of Collective Behavior*,
Nečas Center Series, https://doi.org/10.1007/978-3-030-68147-0_9

Fig. 9.1 Unidirectional
flocks

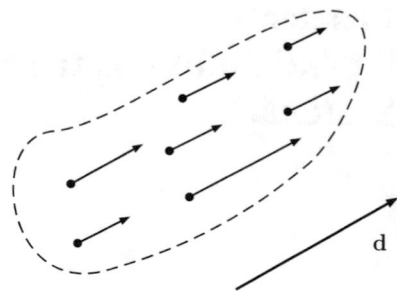

First of all by the maximum principle applied in any direction perpendicular to **d**, one can see that the ansatz (9.1) is preserved in time. Second, in view of rotational invariance of the Euler alignment system, we can assume that **d** points in the direction of the x_1-axis. So, we can assume the horizontal configuration of the flock:

$$\mathbf{u}(x, t) = \langle u(x, t), 0, \ldots, 0 \rangle, \quad u : \mathbb{R}^n \times \mathbb{R}^+ \to \mathbb{R}. \tag{9.2}$$

The full system (6.1) takes the form of a system of scalar conservation laws:

$$(x, t) \in \mathbb{R}^n \times \mathbb{R}^+ \qquad \begin{cases} \partial_t \rho + \partial_1(\rho u) = 0, \\ \partial_t u + \frac{1}{2}\partial_1(u^2) = \phi * (\rho u) - u\phi * \rho. \end{cases} \tag{9.3}$$

The entropy takes form

$$e := \partial_1 u + \phi * \rho, \quad \partial_t e + \partial_1(ue) = 0. \tag{9.4}$$

As a result global well-posedness follows in the same way as in 1D via the threshold condition $e_0 \geq 0$. For solutions with $e_0 > 0$ on Supp ρ_0, we can prove the full analogue of the one-dimensional Theorem 8.1:

$$\mathcal{A}(t) + \|\nabla u(t)\|_{L^\infty(\text{Supp}(\rho(t)))} + \|\nabla^2 u(t)\|_{L^\infty(\text{Supp}(\rho(t)))} \leq Ce^{-\delta t}, \tag{9.5}$$

together with strong flocking (8.6).

We start as before by noting that the diameter of the flock $\mathcal{D}(t)$ remains bounded. So, there exists a time $t^* > 0$ such that

$$e(x, t) \geq \frac{1}{2}\phi(\overline{\mathcal{D}})M, \quad \forall x \in \text{Supp}\,\rho(\cdot, t), \ t > t^*. \tag{9.6}$$

Let us write an equation for $\partial_i u$ in the following form:

$$(\partial_t + u\partial_1)\partial_i u = \mathcal{C}_{\partial_i \phi}(u, \rho) - e\,\partial_i u. \tag{9.7}$$

Recall from Theorem 6.1 that the velocity fluctuations $\mathcal{A}(t)$ are exponentially decaying. Hence, the integral above will be bounded by $\|\partial_i\phi\|_\infty M E(t)$ where we denote by $E(t)$ a generic exponentially decaying quantity. Evaluating (9.7) at the maximum over $\mathrm{Supp}\,\rho(t)$, we obtain

$$\frac{\mathrm{d}}{\mathrm{d}t}\|\partial_i u\|_{L^\infty(\mathrm{Supp}\,\rho(t))} \le E(t) - \tfrac{1}{2}\phi(\overline{\mathcal{D}})M\|\partial_i u\|_{L^\infty(\mathrm{Supp}\,\rho(t))}.$$

This readily implies the exponential bound

$$\|\nabla u(t)\|_{L^\infty(\mathrm{Supp}\,\rho(t))} \le E(t). \tag{9.8}$$

Moving on to the second order derivatives, we write

$$(\partial_t + u\partial_1)\partial_j\partial_i u = \int_\Omega \partial_j\partial_i\phi(x-y)\,(u(y)-u(x))\,\rho(y)\,dy$$

$$- e\,\partial_j\partial_i u - \partial_j e\,\partial_i u - \partial_i e\,\partial_j u.$$

We prove exponential decay by bootstrapping information from the partial derivative $\partial_1\partial_1$, then $\partial_1\partial_j$, and then a general $\partial_i\partial_j$. So, first, we consider the case $i = j = 1$. Note that for this particular case, we have

$$(\partial_t + u\partial_1)\partial_1^2 u = \int_\Omega \partial_1^2\phi(x-y)\,(u(y)-u(x))\,\rho(y)\,dy - e\,\partial_1^2 u - 2\,\partial_1 e\,\partial_1 u.$$

From what we know at this point, the alignment term and $\partial_1 u$ are exponentially decreasing. Using that $\partial_1 e = \partial_1^2 u + \partial_1\phi * \rho$, we arrive at

$$(\partial_t + u\partial_1)\partial_1^2 u = E(t) - e\partial_1^2 u + E(t)\partial_1^2 u.$$

This implies via Rademacher's Lemma

$$\frac{\mathrm{d}}{\mathrm{d}t}\|\partial_1^2 u\|_{L^\infty(\mathrm{Supp}\,\rho(t))} \le E(t) - (\min_{\mathrm{Supp}\,\rho(t)} e)\|\partial_1^2 u\|_{L^\infty(\mathrm{Supp}\,\rho(t))}$$

$$+ E(t)\|\partial_1^2 u\|_{L^\infty(\mathrm{Supp}\,\rho(t))}.$$

Now, as $e(x, t) \ge \tfrac{1}{2}\phi(\overline{\mathcal{D}})M > 0$ for all $x \in \mathrm{Supp}\,\rho(t)$ and $t > t^\star$, we have

$$\frac{\mathrm{d}}{\mathrm{d}t}\|\partial_1^2 u\|_{L^\infty(\mathrm{Supp}\,\rho(t))} \le E(t) - \frac{\mathrm{d}}{\mathrm{d}t}\|\partial_1^2 u\|_{L^\infty(\mathrm{Supp}\,\rho(t))}\left(\tfrac{1}{2}\phi(\overline{\mathcal{D}})M - E(t)\right),$$

for $t > t^\star$. This implies exponential decay by Grönwall's Lemma:

$$\|\partial_1^2 u(t)\|_{L^\infty(\operatorname{Supp}\rho(t))} \le E(t). \tag{9.9}$$

Second, we consider the case $i = 1$ and $j \ne 1$. In this case, we have

$$(\partial_t + u\partial_1)\partial_j\partial_1 u = \int_\Omega \partial_j\partial_1\phi(x - y)\,(u(y) - u(x))\,\rho(y)\,dy - e\,\partial_j\partial_1 u$$
$$- \partial_j e\,\partial_1 u - \partial_1 e\,\partial_j u.$$

Using (9.8) and (9.9), we obtain

$$(\partial_t + u\partial_1)\partial_j\partial_1 u = E(t) - \partial_j\partial_1 u\,(e - E(t)),$$

and by the same argument as above

$$\|\partial_1\partial_j u(t)\|_{L^\infty(\operatorname{Supp}\rho(t))} \le E(t). \tag{9.10}$$

Finally, the case $i, j \ne 1$ relies on the previous ones and proceeds in a similar manner. We have

$$(\partial_t + u\partial_1)\partial_j\partial_i u = E(t) - e\,\partial_j\partial_i u,$$

and hence

$$\|\partial_j\partial_i u(t)\|_{L^\infty(\operatorname{Supp}\rho(t))} \le E(t). \tag{9.11}$$

The same argument as in 1D shows that $\|\nabla\rho\|_\infty$ remains uniformly bounded, and with the exponential decay of the velocity, this implies strong flocking.

Remarkably any 2D-perturbation of a unidirectional flow is still globally well-posed and remains small. This is surprising because such perturbations do not obey conservation law for the e-quantity. Yet, their existence and stability can be established.

Theorem 9.1 *Consider the Euler alignment system* (6.1) *with a smooth absolute kernel ϕ on the periodic domain $\mathbb{T}^n$. Let $(\mathbf{u}_0, \rho_0) \in H^m \times (L_+^1 \cap W^{k,\infty})$ with $m \ge k + 1 > \frac{n}{2} + 2$ and the initial velocity has form*

$$\mathbf{u}_0(x) = u_0(x)\mathbf{d} + v_0(x)\mathbf{d}^\star \qquad \textit{for some } \mathbf{d}, \mathbf{d}^\star \in \mathbb{S}^{n-1}, \tag{9.12}$$

satisfying

$$\inf_{x \in \mathbb{T}^n} e_0(x) \ge \sqrt{\varepsilon}, \quad \|u_0\|_{W^{1,\infty}} \lesssim 1, \quad \|v_0\|_{W^{1,\infty}} \lesssim \varepsilon^2, \tag{9.13}$$

for ε small enough. Then there exists a global in time solution to (6.1) *which is stable around the underlying unidirectional motion:*

$$\|\nabla v(t)\|_\infty \lesssim \varepsilon, \quad \forall t > 0. \tag{9.14}$$

The rest of this section will be devoted to the proof of this statement. The key idea is to analyze an evolution equation for the whole expression on the right-hand side of the e-equation and to establish control over its magnitude which we denote by

$$R(t) := \|(\nabla \cdot \mathbf{u})^2 - \text{Tr}[(\nabla \mathbf{u})^2]\|_\infty.$$

We observe that initially $R(0) \lesssim \varepsilon^2$ and by continuity $R(t) \lesssim \varepsilon^2$ at least for a short period of time. So, let us define a possible critical time t^* at which the solution hypothetically reaches size ε for the first time:

$$R(t^*) = \varepsilon, \qquad R(t) < \varepsilon \quad \text{for } t < t^*.$$

A contradiction will follow if we establish that $R'(t^*) < 0$. This would imply the bound $R(t) < \varepsilon$ on the entire interval of existence, which in turn will imply a bound on e, and since $\phi * \rho$ is bounded a priori, we conclude a bound on $\nabla \cdot \mathbf{u}$. This would imply global existence due to Theorem 7.1.

So, let's write the equation for $(\nabla \cdot \mathbf{u})^2 - \text{Tr}[(\nabla \mathbf{u})^2]$. First, due to rotational invariance of the system (6.1), one can assume for simplicity that $\mathbf{d}$ points in the direction of the x_1-axis and $\mathbf{d}^*$ points in the direction of the x_2-axis:

$$\mathbf{u}_0(x) = \langle u_0(x), v_0(x), 0, \ldots, 0 \rangle. \tag{9.15}$$

Then by the maximum principle, the solution will remain two dimensional for all time:

$$\mathbf{u}(x, t) = \langle u(x, t), v(x, t), 0, \ldots, 0 \rangle. \tag{9.16}$$

Let us define a Poisson-type bracket:

$$\{f, C_\phi(g, h)\} := \partial_1 f \, C_{\partial_2 \phi}(g, h) - \partial_2 f \, C_{\partial_1 \phi}(g, h),$$

and denote

$$\Upsilon_\phi(u, v) := 2 \{u, C_\phi(\rho, v)\} + 2 \{v, C_\phi(\rho, u)\}.$$

Directly from the EAS, we obtain the following equation

$$(\partial_t + \mathbf{u} \cdot \nabla) \left[(\nabla \cdot \mathbf{u})^2 - \text{Tr}[(\nabla \mathbf{u})^2] \right]$$
$$= -(e + (\phi * \rho)) \left[(\nabla \cdot \mathbf{u})^2 - \text{Tr}[(\nabla \mathbf{u})^2] \right] + \Upsilon_\phi(u, v). \tag{9.17}$$

Lemma 9.1 *We have the following bounds on the time interval* $[0, t^*]$:

$$\tfrac{1}{2}\sqrt{\varepsilon} \le e(x, t) \le C_0 := 2\max\{\|e_0\|_\infty, M\|\phi\|_\infty\}. \tag{9.18}$$

Proof Indeed, denote the material derivative by $\dot{e} = (\partial_t + \mathbf{u} \cdot \nabla)e$

$$\dot{e} = \left[(\nabla \cdot \mathbf{u})^2 - \text{Tr}[(\nabla\mathbf{u})^2]\right] + e\left((\phi * \rho) - e\right). \tag{9.19}$$

If $e(x, t) = \tfrac{1}{2}\sqrt{\varepsilon}$ for the first time $t < t^*$, then

$$\dot{e} \ge -\varepsilon + \tfrac{1}{2}\sqrt{\varepsilon}(\phi * \rho(x, t) - \tfrac{1}{2}\sqrt{\varepsilon}) \ge -\varepsilon + \tfrac{1}{2}\sqrt{\varepsilon}(M \inf_x \phi - \tfrac{1}{2}\sqrt{\varepsilon}) > 0,$$

provided $\varepsilon \lesssim M^2$, a contradiction. At the same time, if $e(x, t) = C_0 > \|e_0\|_\infty$, then

$$\dot{e} \le \varepsilon + C_0(M\|\phi\|_\infty - C_0) < 0,$$

provided $C_0 > 2M\|\phi\|_\infty$. So, with the choice of $C_0 = 2\max\{\|e_0\|_\infty, M\|\phi\|_\infty\}$, we have proved the lemma. $\square$

Next, we establish control over the partial gradient $\|\nabla_{1,2}\mathbf{u}(t)\|_\infty$ which is needed to bound the residual term $\Upsilon_\phi(u, v)$ in (9.17). We write an equation for the full gradient in the form suitable for our analysis. We have for the general solutions $\mathbf{u} = (u^1, \ldots, u^n)$

$$(\partial_t + \mathbf{u} \cdot \nabla)\partial_l u^k + (\partial_l\mathbf{u} \cdot \nabla)u^k = \mathcal{C}_{\partial_l\phi}(\rho, u^k) - \partial_l u^k(\phi * \rho).$$

Therefore, we obtain

$$(\partial_t + \mathbf{u} \cdot \nabla)\partial_l u^k = \mathcal{C}_{\partial_l\phi}(\rho, u^k) - \partial_l u^k(\phi * \rho) - (\partial_l\mathbf{w} \cdot \nabla)u^k$$
$$= \mathcal{C}_{\partial_l\phi}(\rho, u^k) - e\,\partial_l u^k + \partial_l u^k(\nabla \cdot \mathbf{u}) - (\partial_l\mathbf{u} \cdot \nabla)u^k,$$

where

$$\partial_l u^k(\nabla \cdot \mathbf{u}) - (\partial_l\mathbf{u} \cdot \nabla)u^k = \sum_{i=1}^n \begin{vmatrix} \partial_i u^i & \partial_l u^i \\ \partial_i u^k & \partial_l u^k \end{vmatrix}.$$

We therefore arrive at

$$(\partial_t + \mathbf{u} \cdot \nabla)\partial_l u^k - \sum_{i=1}^n \begin{vmatrix} \partial_i u^i & \partial_l u^i \\ \partial_i u^k & \partial_l u^k \end{vmatrix} = \mathcal{C}_{\partial_l\phi}(\rho, u^k) - e\,\partial_l u^k. \tag{9.20}$$

For our solutions with only two nonzero components, the gradient takes form

$$\nabla \mathbf{u} = \begin{pmatrix} \partial_1 u \; \partial_2 u & \partial_3 u \ \ldots \ \partial_n u \\ \partial_1 v \; \partial_2 v & \partial_3 v \ \ldots \ \partial_n v \\ \hline 0 \ 0 & 0 \ldots 0 \\ \vdots \ \vdots & \vdots \ \ddots \ \vdots \\ 0 \ 0 & 0 \ldots 0 \end{pmatrix} \equiv \begin{pmatrix} P & Q \\ \hline 0 & 0 \end{pmatrix}.$$

We first address the upper corner block P.

Lemma 9.2 *We have the following bounds on the time interval* $[0, t^*]$

$$\|\partial_1 v(t)\|_\infty + \|\partial_2 v(t)\|_\infty \le C\varepsilon^{3/2}, \qquad \|\partial_1 u(t)\|_\infty + \|\partial_2 u(t)\|_\infty \le C\varepsilon^{-1/2}. \tag{9.21}$$

Proof First, let us consider the off-diagonal entries of P:

$$\begin{cases} (\partial_t + \mathbf{u} \cdot \nabla)\partial_2 u = \mathcal{C}_{\partial_2 \phi}(\rho, u) - e\,\partial_2 u, \\ (\partial_t + \mathbf{u} \cdot \nabla)\partial_1 v = \mathcal{C}_{\partial_1 \phi}(\rho, v) - e\,\partial_1 v. \end{cases} \tag{9.22}$$

In view of (9.18), we have

$$\frac{d}{dt}\|\partial_1 v\|_\infty \le 2M\|\nabla\phi\|_\infty \|v_0\|_\infty - \frac{1}{2}\sqrt{\varepsilon}\,\|\partial_1 v\|_\infty,$$

$$\frac{d}{dt}\|\partial_2 u\|_\infty \le 2M\|\nabla\phi\|_\infty \|u_0\|_\infty - \frac{1}{2}\sqrt{\varepsilon}\,\|\partial_2 u\|_\infty.$$

By Gronwall's inequality

$$\|\partial_1 v\|_\infty \le \|\partial_1 v_0\|_\infty\, e^{-\frac{1}{2}\sqrt{\varepsilon}t} + \frac{2M}{\sqrt{\varepsilon}}\|\partial_1 \phi\|_\infty \|v_0\|_\infty (1 - e^{-\frac{1}{2}\sqrt{\varepsilon}t}) \le C\varepsilon^{3/2},$$

$$\|\partial_2 u\|_\infty \le \|\partial_2 u_0\|_\infty\, e^{-\frac{1}{2}\sqrt{\varepsilon}t} + \frac{2M}{\sqrt{\varepsilon}}\|\partial_2 \phi\|_\infty \|u_0\|_\infty (1 - e^{-\frac{1}{2}\sqrt{\varepsilon}t}) \le C\varepsilon^{-1/2}.$$

As to the diagonal entries of P, we have

$$\begin{cases} (\partial_t + \mathbf{u} \cdot \nabla)\partial_1 u - \frac{1}{2}\left[(\nabla \cdot \mathbf{u})^2 - \mathrm{Tr}[(\nabla\mathbf{u})^2]\right] = \mathcal{C}_{\partial_1 \phi}(\rho, u) - e\,\partial_1 u, \\ (\partial_t + \mathbf{u} \cdot \nabla)\partial_2 v - \frac{1}{2}\left[(\nabla \cdot \mathbf{u})^2 - \mathrm{Tr}[(\nabla\mathbf{u})^2]\right] = \mathcal{C}_{\partial_2 \phi}(\rho, v) - e\,\partial_2 v. \end{cases} \tag{9.23}$$

Thus, on the interval $[0, t^*]$:

$$\frac{d}{dt}\|\partial_1 u\|_\infty \le \varepsilon + 2M\|\partial_1 \phi\|_\infty \|u_0\|_\infty - \frac{1}{2}\sqrt{\varepsilon}\,\|\partial_1 u\|_\infty.$$

Consequently

$$\|\partial_1 u\|_\infty \le \|\partial_1 u_0\|_\infty \, e^{-\frac{1}{2}\sqrt{\varepsilon}t} + \frac{2}{\sqrt{\varepsilon}}\,(M\|\partial_1\phi\|_\infty\|u_0\|_\infty + \varepsilon)(1 - e^{-\frac{1}{2}\sqrt{\varepsilon}t}) \le C\varepsilon^{-1/2}.$$

A similar estimate on $\partial_2 v$ gives us $\lesssim \varepsilon^{1/2}$, which is not sufficient. Instead, we rewrite the equation for $\partial_2 v$ in the following way:

$$(\partial_t + \mathbf{u}\cdot\nabla)\partial_2 v + \partial_2 u\,\partial_1 v + \partial_2 v\partial_2 v = C_{\partial_2\phi}(\rho, v) - (\phi * \rho)\,\partial_2 v. \qquad (9.24)$$

We define the points $x^\pm(t)$ where the maximum and respectively minimum of $\partial_2 v(x, t)$ is attained. Then

$$\frac{d}{dt}\partial_2 v(x^+(t), t) \le -\big[(\phi * \rho) + \partial_2 v(x^+(t), t)\big]\,\partial_2 v(x^+(t), t) + C\varepsilon^{3/2},$$

$$-\frac{d}{dt}\partial_2 v(x^-(t), t) \le \big[(\phi * \rho) + \partial_2 v(x^-(t), t)\big]\,\partial_2 v(x^-(t), t) + C\varepsilon^{3/2}.$$

So, the difference $d(t) := \partial_2 v^\varepsilon(x^+(t), t) - \partial_2 v^\varepsilon(x^-(t), t)$ satisfies

$$d'(t) \le -\big[(\phi * \rho) + \partial_2 v^\varepsilon(x^+(t), t) - \partial_2 v^\varepsilon(x^-(t), t)\big]d(t) + C\varepsilon^{3/2}. \qquad (9.25)$$

We already established that $\|\partial_2 v\|_\infty \le \varepsilon^{1/2}$. Using this we obtain

$$(\phi * \rho) + \partial_2 v^\varepsilon(x^+(t), t) - \partial_2 v^\varepsilon(x^-(t), t) \ge c_0,$$

and hence

$$d'(t) \le -c_0 d(t) + C\varepsilon^{3/2}, \quad d(0) \le \varepsilon^2.$$

Another application of Grönwall's lemma gives $d(t) \lesssim \varepsilon^{3/2}$, and the proof is complete. $\square$

With these ingredients at hand, we now look back at the equation (9.17). At the critical time $t^\star$, we have

$$\frac{d}{dt}R(t^\star) \le -cMR(t^\star) + \Upsilon_\phi(u, v).$$

We have, using (9.21) and $\|v\|_\infty \le \|v_0\|_\infty \approx \varepsilon^2$

$$\{u, C_\phi(\rho, v)\} \le C\varepsilon^{-1/2}\varepsilon^2 = C\varepsilon^{3/2},$$

$$\{v, C_\phi(\rho, u)\} \le C\varepsilon^{3/2}.$$

Consequently

$$\frac{d}{dt}R(t^{\star}) \leq -c\varepsilon + C\varepsilon^{3/2} < 0,$$

for $\varepsilon > 0$ small enough. This shows that $t^* = \infty$, and hence by Lemma 9.1, we have a uniform bound on e, which fulfills the continuation criterion of Theorem 7.1.

Lastly, we establish control over the remaining part Q of the gradient matrix.

Lemma 9.3 *We have for all $k = 3, \ldots, n$ and all time $t > 0$*

$$\|\partial_k u(t)\|_\infty \leq c\varepsilon^{-1/2}, \qquad \|\partial_k v(t)\|_\infty \leq c\varepsilon.$$

Proof Let us write the system for $\partial_k u$ and $\partial_k v$:

$$
\begin{aligned}
(\partial_t + \mathbf{u} \cdot \nabla)\partial_k u - \partial_2 v \, \partial_k u + \partial_2 u \, \partial_k v &= C_{\partial_k\phi}(\rho, u) - e \, \partial_k u, \\
(\partial_t + \mathbf{u} \cdot \nabla)\partial_k v + \partial_1 v \, \partial_k u + \partial_2 v \, \partial_k v &= C_{\partial_k\phi}(\rho, v) - (\phi * \rho) \, \partial_k v.
\end{aligned}
\tag{9.26}
$$

Denote $X_k(t) := \|\partial_k u(t)\|_\infty$ and $Y_k(t) := \|\partial_k v(t)\|_\infty$ for $k = 3, \ldots, n$. Combining (9.21) together with the lower bound $(\phi * \rho) \geq c_0$ and the fact that $e(t) \geq \frac{1}{2}\sqrt{\varepsilon}$, we obtain

$$
\begin{cases}
\dfrac{d}{dt}X_k(t) \leq -c_1\sqrt{\varepsilon}X_k(t) + \dfrac{c_2}{\sqrt{\varepsilon}}Y_k(t) + c_3, \\[2ex]
\dfrac{d}{dt}Y_k(t) \leq c_4\varepsilon^{3/2}X_k(t) - c_5Y_k(t) + c_6\varepsilon^2,
\end{cases}
\tag{9.27}
$$

with all the constants being independent of ε. Defining the vector $Z_k(t) := (X_k(t), Y_k(t))$, the system (9.27) can be rewritten in matrix form as

$$\frac{d}{dt}Z_k(t) \leq AZ_k(t) + b$$

with diagonalization $A = PDP^{-1}$, where

$$A := \begin{pmatrix} -c_1\sqrt{\varepsilon} & \frac{c_2}{\sqrt{\varepsilon}} \\ c_4\varepsilon^{3/2} & -c_5 \end{pmatrix}, \quad b := \begin{pmatrix} c_3 \\ c_6\varepsilon^2 \end{pmatrix} \quad \text{and} \quad D := \begin{pmatrix} \lambda_+(\varepsilon) & 0 \\ 0 & \lambda_-(\varepsilon) \end{pmatrix}.$$

Noting that $\mathrm{Tr}(A) < 0$ and $\mathrm{Det}(A) > 0$ for $\varepsilon > 0$ small enough, both eigenvalues $\lambda_\pm(\varepsilon)$ are negative. More specifically, we have

$$\lambda_\pm(\varepsilon) := \frac{\mathrm{Tr}(A)}{2} \pm \sqrt{\left(\frac{\mathrm{Tr}(A)}{2}\right)^2 - \det(A)}$$

with

$$\lambda_+(\varepsilon) \to -c_5 + \mathcal{O}(\varepsilon) \qquad \text{and} \qquad \lambda_+(\varepsilon) \to -\sqrt{\varepsilon} + \mathcal{O}(\varepsilon).$$

By Duhamel's principle

$$Z_k(t) \leq e^{At} Z_k(0) + \int_0^t e^{A(t-s)} \mathbf{b} \, ds.$$

So, calculating $e^{At} = \mathrm{P} e^{Dt} \mathrm{P}^{-1}$ leads to the solution to the system, by simply integrating with respect to t. Using that $Z_k(0) \approx (1, \varepsilon^2)$ and elementary linear algebra, we obtain

$$X_k(t) \lesssim \left(e^{\lambda_- t} + e^{\lambda_+ t}\right) + \left(\frac{e^{\lambda_- t} - 1}{\lambda_-} + \frac{e^{\lambda_+ t} - 1}{\lambda_+}\right),$$

$$Y_k(t) \lesssim \varepsilon^{3/2} \left(e^{\lambda_- t} - e^{\lambda_+ t}\right) + \varepsilon^{3/2} \left(\frac{e^{\lambda_- t} - 1}{\lambda_-} - \frac{e^{\lambda_+ t} - 1}{\lambda_+}\right).$$

Combining the above we conclude that

$$X_k(t) \leq c_7 \varepsilon^{-1/2}, \quad Y_k(t) \leq c_8 \varepsilon, \quad \forall t \geq 0,$$

as desired.   □

Together with the previously established bound (9.21), we obtain $\|\nabla v(t)\|_\infty \lesssim \varepsilon$. This finishes the proof of Theorem 9.1.

9.2   Mikado Clusters in Hydrodynamic Multi-Flocks

The macroscopic counterpart of the discrete multi-flock system (2.54) introduced in Sect. 2.8 can be derived in a similar fashion to the mono-flock case. Letting macroscopic flock variables denoted by $(\rho_\alpha, \mathbf{u}_\alpha)$ for $\alpha = 1, \ldots, A$ and global flock parameters by

$$\mathbf{X}_\alpha(t) := \frac{1}{M_\alpha} \int_{\mathbb{R}^n} x \, \rho_\alpha(x, t) \, dx,$$

$$M_\alpha := \int_{\mathbb{R}^n} \rho_\alpha(x, t) \, dx,$$

$$\mathbf{V}_\alpha(t) := \frac{1}{M_\alpha} \int_{\mathbb{R}^n} \mathbf{u}_\alpha(x, t) \rho_\alpha(x, t) \, dx,$$

the multi-flock EAS is given by

$$
\begin{cases}
\partial_t \rho_\alpha + \nabla \cdot (\rho_\alpha \mathbf{u}_\alpha) = 0, \\[4pt]
\partial_t \mathbf{u}_\alpha + \mathbf{u}_\alpha \cdot \nabla \mathbf{u}_\alpha = \lambda_\alpha [\phi_\alpha * (\rho_\alpha \mathbf{u}_\alpha) - \mathbf{u}_\alpha (\phi_\alpha * \rho_\alpha)] \\[4pt]
\qquad\qquad + \varepsilon \sum_{\beta \neq \alpha} M_\beta \Psi (\mathbf{X}_\alpha - \mathbf{X}_\beta) (\mathbf{V}_\beta - \mathbf{u}_\alpha),
\end{cases}
\tag{9.28}
$$

where as before communication between flocks is assumed to be weaker than communication inside each of the flocks $\varepsilon \ll \min_\alpha \lambda_\alpha$.

One can develop a similar regularity theory in 1D as for the mono-flock case and show analogues of Theorems 2.7 and 2.8 together with strong flocking statements. Here the α-entropies $e_\alpha = \partial_x u_\alpha + \lambda_\alpha \phi_\alpha * \rho_\alpha$ determine the threshold condition $e_\alpha \geq 0$ for global existence, see [96].

Now, by analogy with the discrete system, we can pass the new system of variables tied to the reference frame of each flock

$$
\mathbf{v}_\alpha(x, t) := \mathbf{u}_\alpha(x - \mathbf{X}_\alpha(t), t) - \mathbf{V}_\alpha(t) \qquad \text{and} \qquad \varrho_\alpha(x, t) := \rho_\alpha(x - \mathbf{X}_\alpha(t), t).
$$

The new system reads

$$
\begin{cases}
\partial_t \varrho_\alpha + \nabla \cdot (\varrho_\alpha \mathbf{v}_\alpha) = 0, \\[4pt]
\partial_t \mathbf{v}_\alpha + \mathbf{v}_\alpha \cdot \nabla \mathbf{v}_\alpha = \lambda_\alpha [\phi_\alpha * (\varrho_\alpha \mathbf{v}_\alpha) - \mathbf{v}_\alpha (\phi_\alpha * \varrho_\alpha)] + \varepsilon R_\alpha(t) \mathbf{v}_\alpha,
\end{cases}
\tag{9.29}
$$

where

$$
R_\alpha(t) := \sum_{\beta \neq \alpha} M_\beta \Psi (\mathbf{X}_\alpha(t) - \mathbf{X}_\beta(t)).
$$

Now, each flock satisfies the maximum principle, and we can study unidirectional configurations:

$$
\mathbf{v}_\alpha(x, t) = v_\alpha(x, t) \mathbf{r}_\alpha \qquad \text{for} \quad v_\alpha : \mathbb{R}^n \times \mathbb{R}^+ \to \mathbb{R}, \quad \mathbf{r}_\alpha \in \mathbb{S}^{n-1}.
\tag{9.30}
$$

We call these solutions Mikado clusters; see Fig. 9.2—by analogy with Mikado solutions to the 3D incompressible Euler equation which played crucial role in the resolution of the celebrated Onsager conjecture [36, 55].

Theorem 9.2 *Consider initial Mikado cluster* (9.30) *with* $(v_\alpha(0), \rho_\alpha(0)) \in H^m \times (L_+^1 \cap W^{1,\infty})$ *with* $m > \frac{n}{2} + 2$, *satisfying the threshold condition* $e_\alpha(0) \geq 0$ *for all* $\alpha = 1, \ldots, A$. *Then there exists a global in time unique solution to system* (9.28) *which retains the form* (9.30) *for all time and lies in the class:*

$$
(v_\alpha, \rho_\alpha) \in C_w([0, \infty); H^m \times (L_+^1 \cap W^{1,\infty})).
$$

Fig. 9.2 Mikado cluster in
v-variables

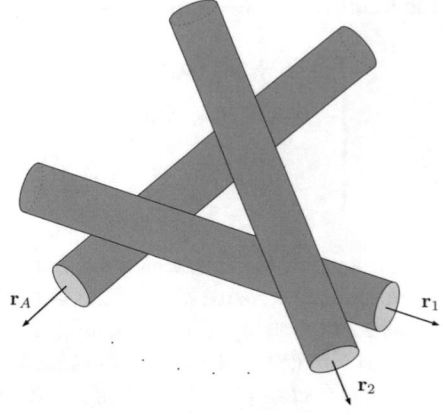

Moreover

Fast local flocking. *Assuming that for a given* $\alpha \in \{1, \ldots, A\}$ *the* α-*flock has
compact support,* $e_\alpha(0) > 0$ *on that support, and the internal kernel* ϕ_α *has
a heavy tail, then there exists a* $\delta_\alpha(\lambda_\alpha, \phi_\alpha, \rho_\alpha(0), \mathbf{u}_\alpha(0))$ *such that*

$$\sup_{x \in \mathrm{Supp}\, \rho_\alpha(\cdot, t)} \left[|\mathbf{u}_\alpha(x, t) - \mathbf{V}_\alpha(t)| + |\nabla \mathbf{u}_\alpha(x, t)| + |\nabla^2 \mathbf{u}_\alpha(x, t)| \right] \lesssim e^{-\delta_\alpha t},$$

$$\|\rho_\alpha(\cdot, t) - \bar\rho_\alpha(\cdot - \mathbf{X}_\alpha(t))\|_{C^\gamma} \lesssim e^{-\delta_\alpha t},$$

for any $0 < \gamma < 1$.

Slow global flocking. *Suppose that the inter-flock kernel* Ψ *has a heavy tail and
the internal kernels* $\phi_\alpha \geq 0$ *are arbitrary. If the multi-flock has a finite diameter
initially, then global alignment occurs at a rate* $\delta(\Psi, \varepsilon, \rho_\alpha(0), \mathbf{u}_\alpha(0))$ *such that*

$$\sup_{\substack{x \in \mathrm{Supp}\, \rho_\alpha(\cdot, t) \\ \alpha = 1, \ldots, A}} \left[|\mathbf{u}_\alpha(x, t) - \mathbf{V}| + |\nabla \mathbf{u}_\alpha(x, t)| + |\nabla^2 \mathbf{u}_\alpha(x, t)| \right] \lesssim e^{-\delta t},$$

$$\|\rho_\alpha(\cdot, t) - \bar\rho_\alpha(\cdot - t\mathbf{V})\|_{C^\gamma} \lesssim e^{-\delta t},$$

where $\mathbf{V} = \frac{1}{M} \sum_{\alpha=1}^{A} M_\alpha \mathbf{V}_\alpha$ *is the global momentum.*

As in the mono-flock case, the e-quantities

$$e_\alpha := \mathbf{r}_\alpha \cdot \nabla v_\alpha + \lambda_\alpha \phi_\alpha * \varrho_\alpha \tag{9.31}$$

play a crucial role. They satisfy a similar conservation law with an extra term

$$\partial_t e_\alpha + \nabla \cdot (\mathbf{v}_\alpha \varrho_\alpha) = -\varepsilon R_\alpha(t) \left(\nabla \cdot \mathbf{v}_\alpha \right),$$

or equivalently, along characteristics

$$\dot{e}_\alpha = (\varepsilon R_\alpha(t) + e_\alpha)(\lambda_\alpha \phi_\alpha * \varrho_\alpha - e_\alpha).$$

Since $R_\alpha \geq 0$, the initial positive entropy $e_\alpha \geq 0$ will preserve its sign, and will also be globally bounded. Thus, $\nabla \cdot \mathbf{u}_\alpha$ is bounded, and hence we obtain global existence by Theorem 7.1.

It was already shown in [96] that any classical solution to a multi-flock aligns exponentially fast. To prove strong flocking, we simply observe that the scalar pair $(v_\alpha, \varrho_\alpha)$ satisfies

$$
\begin{cases}
\partial_t \varrho_\alpha + \nabla \cdot (\varrho_\alpha v_\alpha \mathbf{r}_\alpha) = 0, \\[4pt]
\partial_t v_\alpha + (\mathbf{r}_\alpha \cdot \nabla v_\alpha) \, v_\alpha = \lambda_\alpha \left[\phi_\alpha * (\varrho_\alpha v_\alpha) - v_\alpha (\phi_\alpha * \varrho_\alpha) \right] \\[4pt]
\hspace{3.5cm} + \varepsilon R_\alpha(t) v_\alpha,
\end{cases}
\tag{9.32}
$$

which is similar to (9.3) with the exception of the extra term $\varepsilon R_\alpha(t) v_\alpha$, which is a damping term since $R_\alpha \geq 0$. So, the same analysis as in the previous section applies.

9.3 Spectral Dynamics Approach

In dimension 2 one can obtain an alternative threshold condition based on a spectral dynamics approach. Let us consider the periodic environment, $\mathbb{T}^2$, and assume that the kernel ϕ is smooth, of convolution type, and absolute with $\inf_{\mathbb{T}^2} \phi = \phi_* > 0$. This immediately implies exponential alignment estimate of Theorem 6.1 for any solution on its time interval of existence. Let us assume for simplicity that $\lambda = 1$. Recall that the e-quantity

$$e = \nabla \cdot \mathbf{u} + \phi * \rho, \tag{9.33}$$

satisfies the equation

$$e_t + \nabla \cdot (\mathbf{u}e) = (\nabla \cdot \mathbf{u})^2 - \mathrm{Tr}(\nabla \mathbf{u})^2. \tag{9.34}$$

In 2D the right-hand side is equal exactly to $2 \det(\nabla \mathbf{u})$. So, if we attempt to appeal to the logistic nature of the equation as in 1D, we write (with $\dot{e}$ denoting the material derivative)

$$\dot{e} = e(\phi * \rho - e) + 2 \det(\nabla \mathbf{u}).$$

So, the residual term $\det(\nabla \mathbf{u})$ gets in the way of controlling the growth or the sign of e. It is difficult however to track the dynamics of $\det(\nabla \mathbf{u})$ since $\nabla \mathbf{u}$ is nonsymmetric. Instead, one can track the dynamics of the symmetric part of $\nabla \mathbf{u}$ and in particular the eigenvalues of $S = \frac{1}{2}(\nabla \mathbf{u} + \nabla^\perp \mathbf{u})$. In order to see exactly what we are aiming for, let us note that

$$\det(\nabla \mathbf{u}) = \det S + \omega^2,$$

where $\omega = \frac{1}{2}(\partial_1 u_2 - \partial_2 u_1)$ is the scalar vorticity of the field. Denote by μ_1, μ_2 the eigenvalues of S. Then $\det S = \mu_1 \mu_2$. At the same time

$$\mu_1 \mu_2 = \frac{1}{4}(\mu_1 + \mu_2)^2 - \frac{1}{4}(\mu_1 - \mu_2)^2.$$

The first term is exactly $\frac{1}{4}(\nabla \cdot \mathbf{u})^2$, and the second involves the spectral gap denoted $\eta = \mu_1 - \mu_2$. So

$$2 \det(\nabla \mathbf{u}) = \frac{1}{2}(\nabla \cdot \mathbf{u})^2 - \frac{1}{2}\eta^2 + 2\omega^2.$$

Expanding $\nabla \cdot \mathbf{u} = e - \phi * \rho$, the e-equation can now be rewritten as follows:

$$2\dot{e} = (\phi * \rho)^2 + 4\omega^2 - \eta^2 - e^2. \tag{9.35}$$

The issue now reduces to whether we can control the spectral gap η and the vorticity ω. It turns out that the evolution of both quantities can be read off easily from the equation for $\nabla \mathbf{u}$. Indeed, let us write the full matrix equation first:

$$(\partial_t + \mathbf{u} \cdot \nabla)\nabla \mathbf{u} + (\nabla \mathbf{u})^2 = -(\phi * \rho)\,\nabla \mathbf{u} + E, \tag{9.36}$$

where E is an exponentially decaying quantity. To be precise

$$E = \mathcal{C}_{\nabla \phi}(\mathbf{u}, \rho),$$

and according to Theorem 6.1

$$|E| \le \mathcal{A}_0 e^{-M\phi_* t} \|\nabla \phi\|_\infty M, \tag{9.37}$$

where $\overline{\mathcal{D}}$ is determined solely from the initial condition by equation (2.25). We decompose $\nabla \mathbf{u}$ into its symmetric and skew-symmetric parts:

$$\nabla \mathbf{u} = S + \Omega, \quad S = \frac{1}{2}(\nabla \mathbf{u} + \nabla \mathbf{u}^\top), \quad \Omega = \begin{pmatrix} 0 & -\omega \\ \omega & 0 \end{pmatrix}.$$

Furthermore, we have the following decomposition of the square matrix:

$$(\nabla \mathbf{u})^2 = \underbrace{S^2 - \omega^2 \mathbb{I}}_{\text{sym}} + \underbrace{S\Omega + \Omega S}_{\text{skew-sym}} = S^2 - \omega^2 \mathbb{I} + \Omega \nabla \cdot \mathbf{u}.$$

So, reading off the equation for the skew-symmetric part, we obtain (in Lagrangian coordinates)

$$(\partial_t + \mathbf{u} \cdot \nabla)\omega + e\omega = E.$$

For the symmetric part, we have

$$(\partial_t + \mathbf{u} \cdot \nabla)S + S^2 = \omega^2 \mathbb{I} - (\phi * \rho)\,S + E. \tag{9.38}$$

Now the advantage of considering the symmetric S comes into light at this point. Denote by $(\mathbf{s}_1(x,t), \mathbf{s}_2(x,t))$ the orthonormal basis of eigenvectors corresponding to μ_1 and μ_2, respectively. Then $\mu_i = \mathbf{s}_i\, S \mathbf{s}_i$, and note that

$$(\partial_t + \mathbf{u} \cdot \nabla)\mu_i = \mathbf{s}_i \left[\frac{\mathrm{d}}{\mathrm{d}t} S\right] \mathbf{s}_i,$$

due to orthogonality $\mathbf{s}_i \cdot (\partial_t + \mathbf{u} \cdot \nabla)\mathbf{s}_i = 0$. So, multiplying the S-equation by $\mathbf{s}_i$ from both sides, we obtain a system of equations for the eigenvalues:

$$(\partial_t + \mathbf{u} \cdot \nabla)\mu_i + \mu_i^2 = \omega^2 - (\phi * \rho)\mu_i + E.$$

Lastly, taking the difference and using that $\mu_1^2 - \mu_2^2 = \eta(\nabla \cdot \mathbf{u})$, we obtain

$$(\partial_t + \mathbf{u} \cdot \nabla)\eta + e\eta = E.$$

Collecting the equations, we obtain the system

$$\begin{cases} 2\dot{e} + e^2 = (\phi * \rho)^2 + 4\omega^2 - \eta^2, \\ \dot{\omega} + e\omega = E, \\ \dot{\eta} + e\eta = E. \end{cases} \tag{9.39}$$

Let us note in passing that the bound on E in (9.37) is still valid up to an absolute constant due to the algebraic manipulations above.

So, let us now fix an initial condition $(u_0, \rho_0) \in H^m \times (L_+^1 \cap W^{k,\infty})$ and assume that $e_0(x) > 0$ for every $x \in \mathbb{R}^2$. According to Theorem 7.2, we have a local solution on a maximal time interval $[0, T_0)$. By continuity, $e(X(t,x),t) > 0$ for some short time $t < T(x)$. On that time interval, the spectral gap solution reads

$$\eta(t) = \eta_0 \exp\left\{-\int_0^t e(s)\,\mathrm{d}s\right\} + \int_0^t \exp\left\{-\int_s^t e(\tau)\,\mathrm{d}\tau\right\} E(s)\,\mathrm{d}s.$$

So,

$$|\eta(t)| \le |\eta_0| + \mathcal{A}_0 \frac{\|\nabla \phi\|_\infty}{\phi_*}.$$

Using this we obtain from the e-equation

$$2\dot{e} \geq \phi_*^2 M^2 - \left(|\eta_0| + \mathcal{A}_0 \frac{\|\nabla\phi\|_\infty}{\phi_*}\right)^2 - e^2.$$

Assuming now the small gap and small amplitude conditions

$$|\eta_0| < \frac{1}{4}\phi_* M, \quad \mathcal{A}_0 < \frac{1}{4} \frac{\phi_*^2 M}{\|\nabla\phi\|_\infty} \tag{9.40}$$

we find that

$$2\dot{e} \geq \frac{1}{2}\phi_*^2 M^2 - e^2.$$

This shows that e will remain positive on the entire interval $[0, T_0)$, which implies that $\nabla \cdot \mathbf{u}$ remains bounded from below. The continuation criterion (7.9) proves global existence.

Theorem 9.3 *Suppose that $(u_0, \rho_0) \in H^m \times (L_+^1 \cap W^{1,\infty})$, and assume that $e_0(x) > 0$ for every $x \in \mathbb{T}^2$. Assume also the smallness conditions* (9.40). *Then there exists a global solution with this initial condition.*

It is interesting to note that the small amplitude condition alone would guarantee global existence for singular models due to an additional dissipation enhancement effect, see Sect. 9.4.

Let us observe that in the case of unidirectional flocks described in the previous section, $|\eta_0| = |\nabla\mathbf{u}_0|$. So, the spectral gap does not always apply to those flows. Yet, we know that they are globally well-posed. It would be interesting to bridge the gap between the two classes of solutions.

9.4   Nearly Aligned Flocks of Singular Models: Small Initial Data

The lack of control on e in the multidimensional case is part of the reason why the model has only partially developed well-posedness theory. For singular models, however, dissipation provided by the alignment may be strong enough for some solutions to preserve their regularity. In this section we uncover one such class— solutions with a small velocity amplitude $\mathcal{A}_0$, or nearly aligned flocks.

To fix the notation, we denote by $[\,\cdot\,]_s$ the metric of the homogeneous Hölder class $C^s(\mathbb{T}^n)$. For higher s, we will resort to a finite difference definition of $[\,\cdot\,]_s$ stated as follows. First we denote

$$\delta_h f(x) = f(x+h) - f(x), \quad \tau_z f(x) = f(x+z)$$
$$\delta_h^2 f = \delta_h(\delta_h f), \quad \delta_h^3 f = \delta_h(\delta_h(\delta_h f)).$$

We then define, for $0 < \gamma < 1$:

$$[f]_{2+\gamma} = \sup_{x,h \in \mathbb{T}^n} \frac{|\delta_h^3 f(x)|}{|h|^{2+\gamma}}. \tag{9.41}$$

The equivalence of (9.41) to the classical norm $[\nabla^2 f]_\gamma$ is a well-known result in approximation theory; see [101]. For integer values of the smoothness parameter $k \in \mathbb{N}$, we use the classical homogeneous metric $[f]_k = \|\nabla^k f\|_\infty$.

We consider the global singular communication given by the kernel of the classical fractional Laplacian $\phi = \phi_\alpha$, see (7.14)–(7.16).

Let us state the main result.

Theorem 9.4 *Consider the Euler alignment system (6.1) on the torus $\mathbb{T}^n$ with singular kernel given by (7.14), $0 < \alpha < 2$. There exists an $N \in \mathbb{N}$ such that for any sufficiently large $R > 0$, depending only on α and the dimension n, any initial condition $(\mathbf{u}_0, \rho_0) \in H^{m+1}(\mathbb{T}^n) \times H^{m+\alpha}(\mathbb{T}^n)$, $m > \frac{n}{2} + 3$, satisfying*

$$\|\rho_0\|_\infty, \|\rho_0^{-1}\|_\infty, [\mathbf{u}_0]_3, [\rho_0]_3 \leq R,$$

$$\mathcal{A}_0 \leq \frac{1}{R^N}, \tag{9.42}$$

gives rise to a unique global solution in the class

$$(\mathbf{u}, \rho) \in C_w([0, \infty); H^{m+1} \times H^{m+\alpha}).$$

Moreover, the solution converges to a flocking state exponentially fast:

$$\mathcal{A}(t) + [\mathbf{u}(t)]_1 + [\mathbf{u}(t)]_2 + \|\rho(t) - \bar{\rho}(t)\|_{C^1} < Ce^{-\delta t}.$$

In the course of the proof of Theorem 9.4, we establish uniform control on the C^2-norm of $\mathbf{u}$ and the distance between the initial density ρ_0 and its final profile $\bar{\rho}$. As a byproduct, we obtain the following stability result for flocking states.

Theorem 9.5 *Let $(\bar{\mathbf{u}}, \bar{\rho})$ be a traveling wave, where $\bar{\rho}(x, t) = \bar{\rho}(x - t\bar{\mathbf{u}})$, and let $(\mathbf{v}_0, r_0)$ be an initial data satisfying the conditions of Theorem 9.4. Suppose $\|\mathbf{v}_0 - \bar{\mathbf{u}}\|_\infty + \|r_0 - \bar{\rho}_0\|_\infty < \varepsilon$. Then the solution will converge to another flock $\bar{r}$ with $\|\bar{r} - \bar{\rho}_0\|_\infty < \varepsilon^\theta$, where $\theta \in (0, 1)$ depends only on α.*

The idea of the proof is to establish control over a higher Hölder norm $[\mathbf{u}]_{2+\gamma}$. This serves multiple purposes. First, it automatically shows boundedness of the gradients $\|\nabla \mathbf{u}\|_\infty$ and $\|\nabla \rho\|_\infty$, where for the latter we need to control $\|\nabla^2 \mathbf{u}\|_\infty$. So, we fulfill the continuation criterion of Theorem 7.4 and conclude global existence as stated. Second, with the $C^{2+\gamma}$ norm uniformly bounded, we obtain exponential decay of $[\mathbf{u}(t)]_1 + [\mathbf{u}(t)]_2$ simply by interpolation with $\mathcal{A}$, which readily implies strong flocking as in Theorem 8.1.

From now on we will fix an exponent $0 < \gamma < 1$ to be identified later but dependent only on α.

The proof will be structured in several steps.

Step 1: Breakthrough Scenario According to Theorem 7.4, we have a local solution $(\mathbf{u}, \rho) \in C_w([0, T_0) : H^{m+1} \times H^{m+\alpha})$ satisfying the assumptions of Theorem 9.4. Note that in view of the smallness assumption on $\mathcal{A}_0$, the norm $[\mathbf{u}(t)]_{2+\gamma}$ will remain smaller than 1 at least for a short period of time. If the solution cannot be extended beyond T_0, there exists a possible critical time $t^* < T_0$ at which the solution reaches size R for the first time:

$$[\mathbf{u}(t^*)]_{2+\gamma} = R, \quad [\mathbf{u}(t)]_{2+\gamma} < R, \quad t < t^*. \tag{9.43}$$

A contradiction is achieved when we show that $\partial_t [\mathbf{u}(t^*)]_{2+\gamma} < 0$. This establishes the bound $[\mathbf{u}(t)]_{2+\gamma} < R$ on the entire interval $[0, T_0)$ and, hence, continuation of the solution beyond T_0 by Theorem 7.4. In the course of the argument, we pick γ based on several restrictions that occur in the course of the argument but ultimately depending only on α.

Step 2: Preliminary Estimates on $[0, t^*]$ We will derive a few preliminary estimates on various Hölder norms of the data. We fix R and N that are sufficiently large, and N depending only on α, for all the arguments below to go through.

First, we notice two direct bounds:

$$[\mathbf{u}(t)]_1, [\mathbf{u}(t)]_2 < R^{-\frac{6}{\alpha}} e^{-\frac{c_0 t}{R}}, \quad \text{for all } t \le t^*. \tag{9.44}$$

Indeed, by interpolation and in view of (9.42)

$$[\mathbf{u}]_1 \le \mathcal{A}_0^{\frac{1+\gamma}{2+\gamma}} [\mathbf{u}]_{2+\gamma}^{\frac{1}{2+\gamma}} \le R^{1-N/2} e^{-c_0 t/R} < R^{-\frac{6}{\alpha}} e^{-\frac{c_0 t}{R}},$$

and similarly

$$[\mathbf{u}]_2 \le \mathcal{A}_0^{\frac{\gamma}{2+\gamma}} [\mathbf{u}]_{2+\gamma}^{\frac{2}{2+\gamma}} < R^{1-N\frac{\gamma}{2+\gamma}} e^{-c_0 t/R} \le R^{-\frac{6}{\alpha}} e^{-c_0 t/R}.$$

Next, we consider the density. Let us denote $\underline{\rho}$ and $\overline{\rho}$ the minimum and maximum of ρ, respectively. From (7.7) we conclude the bounds

$$\underline{\rho}_0 \exp\left\{ -\int_0^t \|\nabla \cdot \mathbf{u}(s)\|_\infty \, ds \right\} \le \underline{\rho}(t), \quad \overline{\rho}(t) \le \overline{\rho}_0 \exp\left\{ \int_0^t \|\nabla \cdot \mathbf{u}(s)\|_\infty \, ds \right\}.$$

By (9.44), $\|\nabla \cdot \mathbf{u}(s)\|_\infty \le R^{-3} e^{-c_0 s/R}$. Consequently

$$\int_0^t \|\nabla \cdot \mathbf{u}(s)\|_\infty \, ds \le c R^{-2} \le \ln 2.$$

We have arrived at

$$\frac{1}{2R} \leq \underline{\rho}(t), \quad \overline{\rho}(t) \leq 2R. \tag{9.45}$$

Next we obtain higher-order bounds on ρ with the help of the e-quantity. Note that the right-hand side of the e-equation (7.22) is bounded by

$$|(\nabla \cdot \mathbf{u})^2 - \mathrm{Tr}(\nabla \mathbf{u})^2| \leq c[\mathbf{u}]_1^2 \lesssim R^{-6} e^{-c_0 t/R}.$$

So, from (7.22) we obtain

$$\frac{\mathrm{d}}{\mathrm{d}t} \|e\|_\infty \leq R^{-3} e^{-c_0 t/R} \|e\|_\infty + R^{-6} e^{-c_0 t/R}.$$

By Grönwall's Lemma, and using that $\|e_0\|_\infty < R$, we conclude that

$$\|e(t)\|_\infty \leq 2R, \quad t < t^*. \tag{9.46}$$

By a similar computation for ∇e, we obtain

$$\frac{\mathrm{d}}{\mathrm{d}t} [e]_1 \lesssim [\mathbf{u}]_1 [e]_1 + [\mathbf{u}]_2 \|e\|_\infty + c[\mathbf{u}]_1 [\mathbf{u}]_2,$$

and using (9.44)

$$\frac{\mathrm{d}}{\mathrm{d}t} [e]_1 \lesssim R^{-3} e^{-c_0 t/R} [e]_1 + 2R^{-2} e^{-c_0 t/R} + cR^{-3} e^{-c_0 t/R}.$$

Since initially $[e_0]_1 \leq [\mathbf{u}_0]_2 + [\rho_0]_3 < 2R$, again by Grönwall's Lemma

$$[e]_1 \leq 4R.$$

This implies that $[\Lambda_\alpha \rho]_1 < 5R$. So, if $\alpha \neq 1$, this translates directly into the Hölder norm, and we obtain

$$[\rho]_{1+\alpha} \leq c_0 R, \tag{9.47}$$

while for $\alpha = 1$; however it implies bounds in other borderline classes, and as a consequence

$$[\rho]_{2-\gamma} \leq c_0 R. \tag{9.48}$$

Step 3: Higher-Order Nonlocal Maximum Principle Here we adapt the argument of Lemma 8.6 to obtain a nonlocal maximum principle for higher-order finite differences. As before we denote

$$D_\alpha f(x) = \int_{\mathbb{R}^n} |f(x+z) - f(x)|^2 \frac{dz}{|z|^{n+\alpha}}. \tag{9.49}$$

Lemma 9.4 *There is an absolute constant $c_0 > 0$ such that*

$$D_\alpha[\delta_h^3 f](x) \geq c_0 \frac{|\delta_h^3 f(x)|^{2+\alpha}}{[f]_2^\alpha |h|^{3\alpha}}. \tag{9.50}$$

Proof Fix a smooth cutoff function χ, and $r > 0$ to be specified later. We obtain

$$D_\alpha[\delta_h^3 f](x) \geq \int_{\mathbb{R}^n} |\delta_z \delta_h^3 f(x)|^2 \frac{1 - \chi(z/r)}{|z|^{n+\alpha}} \, dz$$

$$\geq \int_{\mathbb{R}^n} (|\delta_h^3 f(x)|^2 - 2\delta_h^3 f(x)\delta_h^3 f(x+z)) \frac{1 - \chi(z/r)}{|z|^{n+\alpha}} \, dz$$

$$\geq |\delta_h^3 f(x)|^2 \frac{1}{r^\alpha} - 2\delta_h^3 f(x) \int_{\mathbb{R}^n} \delta_h^3 f(x+z) \frac{1 - \chi(z/r)}{|z|^{n+\alpha}} \, dz.$$

Note the Taylor residue formula

$$\delta_h^3 f(x+z) = \int_0^1 \int_0^1 \int_0^1 \nabla_z^3 f(x+z+(\theta_1+\theta_2+\theta_3)h)(h, h, h) \, d\theta_1 \, d\theta_2 \, d\theta_3.$$

Integrating by parts in z, and using the bound

$$\left| \nabla_z \left(\frac{1 - \chi(z/r)}{|z|^{n+\alpha}} \right) \right| \leq \frac{c}{|z|^{n+\alpha+1}} \chi_{|z|>r},$$

we obtain

$$\left| \int_{\mathbb{R}^n} \delta_h^3 f(x+z) \frac{1 - \chi(z/r)}{|z|^{n+\alpha}} \, dz \right| \leq C[f]_2 \frac{|h|^3}{r^{\alpha+1}}.$$

Continuing with the main estimate, we obtain

$$D_\alpha[\delta_h^3 f](x) \geq |\delta_h^3 f(x)|^2 \frac{1}{r^\alpha} - C[f]_2 |\delta_h^3 f(x)| \frac{|h|^3}{r^{\alpha+1}}.$$

Choosing $r \sim \frac{[f]_2 |\delta_h^3 f(x)||h|^3}{|\delta_h^3 f(x)|^2}$ produces (9.50). $\qquad \square$

Step 4: Main Estimates We are now in a position to use the velocity equation to derive estimates on the derivative of $[\mathbf{u}]_{2+\gamma}$. Let us fix a pair $(x, h) \in \mathbb{T}^n$ which maximizes (9.41), and we have at time t^*

$$[\mathbf{u}]_{2+\gamma} = \frac{|\delta_h^3 \mathbf{u}(x)|}{|h|^{2+\gamma}} = R.$$

Consequently, at this time, using (9.50)

$$\frac{1}{|h|^{4+2\gamma}} D_\alpha \delta_h^3 \mathbf{u}(x) \geq \frac{R^{8+\alpha}}{|h|^{\alpha(1-\gamma)}}. \tag{9.51}$$

Let us now write the equation for the third finite difference:

$$\partial_t \delta_h^3 \mathbf{u} + \delta_h^3 (\mathbf{u} \cdot \nabla \mathbf{u}) = \int_{\mathbb{R}^n} \delta_h^3 [\rho(x+z)\delta_z \mathbf{u}(x)] \frac{dz}{|z|^{n+\alpha}}. \tag{9.52}$$

Let us denote the transport and alignment terms by

$$B = \delta_h^3 (\mathbf{u} \cdot \nabla \mathbf{u}),$$
$$I = \int_{\mathbb{R}^n} \delta_h^3 [\rho(x+z)\delta_z \mathbf{u}(x)] \frac{dz}{|z|^{n+\alpha}}. \tag{9.53}$$

Let use the test-function $\delta_h^3 \mathbf{u}(x)/|h|^{4+2\gamma}$ and evaluate at the maximizing pair at which point we also have

$$\frac{\delta_h^3 \mathbf{u}(x)}{|h|^{4+2\gamma}} = \frac{[\mathbf{u}]_{2+\gamma}}{|h|^{2+\gamma}}.$$

So we obtain from the equation

$$\partial_t [\mathbf{u}]_{2+\gamma}^2 + \frac{[\mathbf{u}]_{2+\gamma}}{|h|^{2+\gamma}} B = \frac{\delta_h^3 \mathbf{u}(x)}{|h|^{4+2\gamma}} I.$$

Let us first estimate the transport term. By the product formula

$$\delta_h^3 (fg) = \delta_h^3 f \tau_{3h} g + 3\delta_h^2 f \delta_h \tau_{2h} g + 3\delta_h f \delta_h^2 \tau_h g + f \delta_h^3 g.$$

Thus

$$B = \delta_h^3 \mathbf{u} \cdot \tau_{3h} \nabla \mathbf{u} + 3\delta_h^2 \mathbf{u} \cdot \delta_h \tau_{2h} \nabla \mathbf{u} + 3\delta_h \mathbf{u} \cdot \delta_h^2 \tau_h \nabla \mathbf{u} + \mathbf{u} \cdot \nabla \delta_h^3 \mathbf{u}.$$

Note that the last term vanishes due to criticality. Consequently

$$\frac{1}{|h|^{2+\gamma}} |B| \leq [\mathbf{u}]_{2+\gamma} [\mathbf{u}]_1 + 3|h|^{1-\gamma} [\mathbf{u}]_2^2 + 3[\mathbf{u}]_1 [\mathbf{u}]_{2+\gamma} \lesssim [\mathbf{u}]_{2+\gamma} [\mathbf{u}]_1 + |h|^{1-\gamma} [\mathbf{u}]_2^2.$$

Multiplying by another $[\mathbf{u}]_{2+\gamma} \leq R$ and using (9.44), we obtain

$$\frac{[\mathbf{u}]_{2+\gamma}}{|h|^{2+\gamma}}|B| \lesssim R^{-1} + R^{-5} < 1. \tag{9.54}$$

We now turn to the dissipation term. The integrand is given by $\delta_h^3[\tau_z \rho \, \delta_z \mathbf{u}]$. So, we expand by the product rule and using commutativity $\delta_h \delta_z = \delta_z \delta_h$:

$$\delta_h^3[\tau_z \rho \, \delta_z \mathbf{u}] = \delta_h^3 \tau_z \rho \, \tau_{3h} \delta_z \mathbf{u} + 3\delta_h^2 \tau_z \rho \, \tau_{2h} \delta_h \delta_z \mathbf{u} + 3\delta_h \tau_z \rho \, \tau_h \delta_h^2 \delta_z \mathbf{u} + \tau_z \rho \, \delta_z \delta_h^3 \mathbf{u}.$$

Multiplying by $\delta_h^3 \mathbf{u}$, the last term provides the necessary dissipation:

$$\tau_z \rho \, \delta_z \delta_h^3 \mathbf{u} \, \delta_h^3 \mathbf{u} \le -\frac{1}{2}\rho \, |\delta_z \delta_h^3 \mathbf{u}|^2.$$

Dividing by $|h|^{4+2\gamma}$ and using (9.51), we obtain the lower bound

$$\frac{1}{2|h|^{4+2\gamma}}\rho D_\alpha \delta_h^3 \mathbf{u}(x) \ge \frac{R^7}{|h|^{\alpha(1-\gamma)}}. \tag{9.55}$$

In particular we can see that the entire transport term estimated in (9.54) is absorbed by the dissipation at the time t^*. As a result we obtain the inequality

$$\partial_t [\mathbf{u}]_{2+\gamma}^2 \le -\frac{R^7}{|h|^{\alpha(1-\gamma)}} + \frac{\delta_h^3 \mathbf{u}(x)}{|h|^{4+2\gamma}} J,$$

where J contains all the remaining three terms of I:

$$J = \int_{\mathbb{R}^n} [\delta_h^3 \tau_z \rho \, \tau_{3h} \delta_z \mathbf{u} + 3\delta_h^2 \tau_z \rho \, \tau_{2h} \delta_h \delta_z \mathbf{u} + 3\delta_h \tau_z \rho \, \tau_h \delta_h^2 \delta_z \mathbf{u}] \frac{dz}{|z|^{n+\alpha}}$$
$$= J_1 + 3J_2 + 3J_3.$$

For the remainder of the proof, we provide estimates for each of the J_i terms with the common goal to obtain the bound

$$\frac{1}{|h|^{2+\gamma}}|J_i| \lesssim \frac{|h|^\varepsilon}{|h|^{\alpha(1-\gamma)}}, \tag{9.56}$$

for some $\varepsilon > 0$ and γ is sufficiently small. If this is achieved, then the dissipation absorbs all these remaining J-terms, and we conclude that

$$\partial_t [\mathbf{u}(t^*)]_{2+\gamma}^2 < 0,$$

which would finish the proof.

So, let us begin with J_1. Symmetrizing in z we obtain

$$J_1 = \frac{1}{2} \int_{\mathbb{R}^n} [\delta_h^3(\tau_z \rho - \tau_{-z} \rho) \, \tau_{3h} \delta_z \mathbf{u} + \delta_h^3 \tau_z \rho \, \tau_{3h}(\delta_z \mathbf{u} + \delta_{-z} \mathbf{u})] \frac{dz}{|z|^{n+\alpha}}. \qquad (9.57)$$

For the first summand we use

$$|\delta_h^3(\tau_z \rho - \tau_{-z} \rho) \, \tau_{3h} \delta_z \mathbf{u}| \leq |h|^{\alpha - \gamma} \min\{|z|^2, |h|\}.$$

For the second summand

$$|\delta_h^3 \tau_z \rho \, \tau_{3h}(\delta_z \mathbf{u} + \delta_{-z} \mathbf{u})| \leq |h|^{1+\alpha - \gamma} \min\{|z|^2, 1\}.$$

Plugging into (9.57) and integrating, we obtain the desired bound (9.56), but the computation extends only up to $\alpha > \frac{1}{2}$. The problem is that the density receives all the δ_h's and does not fully utilize them. At the same time, $\mathbf{u}$ can no longer directly contribute powers of h. So, we switch one h-difference back onto $\mathbf{u}$. So, let us fix $0 < \alpha \leq \frac{1}{2}$. We start from the original formula:

$$J_1 = \int_{\mathbb{R}^n} \delta_h^3 \tau_z \rho(x) \, \tau_{3h} \delta_z \mathbf{u}(x) \frac{dz}{|z|^{n+\alpha}}.$$

Over the domain $|z| < 10|h|$, we estimate using a cutoff function χ as before

$$\int_{\mathbb{R}^n} |\delta_h^3 \tau_z \rho(x) \, \tau_{3h} \delta_z \mathbf{u}(x)| \, \chi\left(\frac{z}{10|h|}\right) \frac{dz}{|z|^{n+\alpha}} \leq \int_{|z|<10|h|} |h|^{1+\alpha} \frac{dz}{|z|^{n+\alpha-1}} \lesssim |h|^2.$$

This culminates in (9.56). For the remaining part, denote for clarity $f = \delta_h^2 \rho$. So, $\delta_h^3 \tau_z \rho(x) = f(x+h+z) - f(x+z)$. Let us write

$$\int_{\mathbb{R}^n} (f(x+h+z) - f(x+z)) \, \tau_{3h} \delta_z \mathbf{u}(x) \frac{(1 - \chi(\frac{z}{10|h|})) \, dz}{|z|^{n+\alpha}}$$

$$= \int_{\mathbb{R}^n} f(x+z) \left(\tau_{3h} \delta_{z-h} \mathbf{u}(x) \frac{(1 - \chi(\frac{z-h}{10|h|}))}{|z-h|^{n+\alpha}} - \tau_{3h} \delta_z \mathbf{u}(x) \frac{(1 - \chi(\frac{z}{10|h|}))}{|z|^{n+\alpha}} \right) dz$$

$$= \int_{\mathbb{R}^n} f(x+z) \tau_{3h}(\delta_{z-h} \mathbf{u}(x) - \delta_z \mathbf{u}(x)) \frac{(1 - \chi(\frac{z-h}{10|h|}))}{|z-h|^{n+\alpha}} \, dz$$

$$- \int_{\mathbb{R}^n} f(x+z) \tau_{3h} \delta_z \mathbf{u}(x) \left(\frac{(1 - \chi(\frac{z-h}{10|h|}))}{|z-h|^{n+\alpha}} - \frac{(1 - \chi(\frac{z}{10|h|}))}{|z|^{n+\alpha}} \right) dz.$$

All the integrals are supported on $|z| > 9|h|$, where $|z - h| \sim |z|$. Estimating the first one, we use

$$|\delta_{z-h}\mathbf{u}(x) - \delta_z\mathbf{u}(x)| = |\mathbf{u}(x+z-h) - \mathbf{u}(x+z)| \le |h|$$

$$|f(x+z)| \le |h|^{1+\alpha}.$$

Consequently

$$\left| \int_{\mathbb{R}^n} f(x+z) \tau_{3h}(\delta_{z-h}\mathbf{u}(x) - \delta_z\mathbf{u}(x)) \frac{(1 - \psi(\frac{z-h}{10|h|}))}{|z-h|^{n+\alpha}} \, dz \right|$$

$$\le |h|^{2+\alpha} \int_{|z| \ge |h|} \frac{dz}{|z|^{n+\alpha}} \le |h|^2,$$

which implies (9.56). For the second integral, we use

$$\left| \frac{(1 - \psi(\frac{z-h}{10|h|}))}{|z-h|^{n+\alpha}} - \frac{(1 - \psi(\frac{z}{10|h|}))}{|z|^{n+\alpha}} \right| \le \frac{|h|}{|z - \theta h|^{n+\alpha+1}} \mathbb{1}_{|z|>9|h|}$$

$$\lesssim \frac{|h|}{|z|^{n+\alpha+1}} \mathbb{1}_{|z|>9|h|},$$

and

$$|f(x+z)\tau_{3h}\delta_z\mathbf{u}(x)| \le |h|^{1+\alpha}|z|.$$

Integration reproduces the same bound as for the first part.

Next, J_2. For $\alpha < 1$, let us use (9.44) and (9.47) to deduce

$$|\delta_h^2 \tau_z \rho| \le [\rho]_{1+\alpha} |h|^{1+\alpha} \lesssim R|h|^{1+\alpha},$$

$$|\tau_{2h}\delta_h\delta_z\mathbf{u}| \le [\mathbf{u}]_2 |h| \min\{|z|, 1\} \lesssim R^{-1}|h| \min\{|z|, 1\}.$$

The singularity is now removed, and we get

$$\frac{1}{|h|^{2+\gamma}}|J_2| \lesssim |h|^{\alpha-\gamma},$$

which implies (9.56) for sufficiently small γ. For $\alpha \ge 1$ we first symmetrize

$$J_2 = \frac{1}{2} \int_{\mathbb{R}^n} [\delta_h^2(\tau_z - \tau_{-z})\rho \, \tau_{2h}\delta_h\delta_z\mathbf{u} + \delta_h^2\tau_z\rho \, \tau_{2h}\delta_h(\delta_z + \delta_{-z})\mathbf{u}] \frac{dz}{|z|^{n+\alpha}}.$$

Here for the first summand, we use (9.48):

$$|\delta_h^2(\tau_z - \tau_{-z})\rho| \le R \min\{|h|^{1+\alpha-\gamma}, |h|^{\alpha-\gamma}|z|\},$$

$$|\tau_{2h}\delta_h\delta_z\mathbf{u}| \le R^{-1}|h| \min\{|z|, 1\}.$$

With this at hand, we proceed

$$\frac{1}{|h|^{2+\gamma}} \int_{\mathbb{R}^n} |\delta_h^2(\tau_z - \tau_{-z})\rho \, \tau_{2h}\delta_h\delta_z \mathbf{u}| \frac{dz}{|z|^{n+\alpha}} \leq \frac{|h|^{1+\alpha-\gamma}}{|h|^{2+\gamma}} \leq \frac{|h|^{\alpha-2\gamma-1+\alpha(1-\gamma)}}{|h|^{\alpha(1-\gamma)}}$$
$$\leq \frac{|h|^{\varepsilon}}{|h|^{\alpha(1-\gamma)}},$$

since clearly, $\alpha - 2\gamma - 1 + \alpha(1-\gamma) > 0$ for small g. In the second summand, using that $(\delta_z + \delta_{-z})\mathbf{u}$ is the second order finite difference

$$|\delta_h^2 \tau_z \rho \, \tau_{2h}\delta_h(\delta_z + \delta_{-z})\mathbf{u}| \leq |h|^{2-\gamma} \min\{|z|^2, 1\},$$

we obtain

$$\frac{1}{|h|^{2+\gamma}} \int_{\mathbb{R}^n} |\delta_h^2 \tau_z \rho \, \tau_{2h}\delta_h(\delta_z + \delta_{-z})\mathbf{u}| \frac{dz}{|z|^{n+\alpha}} \leq \frac{|h|^{2-\gamma}}{|h|^{2+\gamma}} \leq \frac{h^{\alpha(1-\gamma)-2\gamma}}{|h|^{\alpha(1-\gamma)}}.$$

This finishes the bound on J_2.

Finally, for J_3 we proceed similarly. For $\alpha < 1$, we use

$$|\delta_h \tau_z \rho \, \tau_h \delta_h^2 \delta_z \mathbf{u}| \leq |h|^2 \min\{|z|, 1\}.$$

Hence

$$\frac{1}{|h|^{2+\gamma}} |J_3| \lesssim \frac{|h|^2}{|h|^{2+\gamma}} \leq \frac{h^{\alpha(1-\gamma)-\gamma}}{|h|^{\alpha(1-\gamma)}} \leq \frac{h^{\varepsilon}}{|h|^{\alpha(1-\gamma)}}.$$

For $\alpha \geq 1$, we again symmetrize first

$$J_3 = \frac{1}{2} \int_{\mathbb{R}^n} [\delta_h(\tau_z - \tau_{-z})\rho \, \tau_h \delta_h^2 \delta_z \mathbf{u} + \delta_h \tau_z \rho \, \tau_h \delta_h^2(\delta_z + \delta_{-z})\mathbf{u}] \frac{dz}{|z|^{n+\alpha}},$$

and using

$$|\delta_h(\tau_z - \tau_{-z})\rho \, \tau_h \delta_h^2 \delta_z \mathbf{u}| \leq \min\{|h|^{3+\alpha-\gamma}, |h|^{1+\alpha}|z|^2\},$$
$$|\delta_h \tau_z \rho \, \tau_h \delta_h^2(\delta_z + \delta_{-z})\mathbf{u}| \leq \min\{|h|^3, |h||z|^2\},$$

integration implies (9.56).

We have established that $\partial_t[\mathbf{u}(t^*)]_{2+\gamma}^2 < 0$ at the critical time. This means that such time t^* does not exist and finishes the proof of the existence part.

Step 5: Flocking and Stability As we noted in the beginning, exponential decay of $[\mathbf{u}]_2^2$ implies uniform control over $\|\nabla\rho\|_\infty$.

Arguing as in the proof of Theorem 8.1, we can slightly improve the space in which strong flocking occurs. This is due to (9.47)–(9.48) bounds, which imply that

$\bar{\rho} \in W^{1+\alpha-\gamma,\infty}$ by compactness. Using again (9.48) and by interpolation, we have convergence in the $W^{1,\infty}$-metric as well:

$$[\rho(\cdot, t) - \bar{\rho}(\cdot, t)]_1 < C_2 e^{-\delta t}.$$

As far as stability is concerned, the computation above shows that in fact the limiting profile $\bar{r}$ differs little from the initial density r_0 under the conditions of Theorem 9.5. Indeed, setting R such that $\varepsilon = 1/R^N$ (here $\varepsilon > 0$ is small), we obtain via (9.44)

$$\|\partial_t r\|_\infty \leq C R^{-2} e^{-c_0 t/R}.$$

Hence, $\|\bar{r} - r_0\|_\infty \leq \frac{C}{c_0 R} = \varepsilon^\theta$. Since $\|r_0 - \bar{\rho}_0\|_\infty < \varepsilon$, this finishes the result.

9.5   Notes and References

The results of Sects. 9.1 and 9.2 are taken from an upcoming publication [65]; see also [66] for an extension to singular models. The spectral dynamics approach was initially proposed in Tadmor and Tan [99] and reached its present form in He and Tadmor [52]. The work [99] also examines various threshold conditions of regularity in the 2D setting. It remains open whether there is an umbrella family of solutions that would incorporate both unidirectional and small spectral gap classes.

The results of Sect. 9.4 were proved in [92]. An alternative approach appeared in Danchin et al. [35] with a smallness condition formulated in Besov spaces with positive smoothness. Due to lack of a good continuation criterion, it is not clear whether similar results hold for topological models.

References

1. Albi G, Bellomo N, Fermo L, Ha S-Y, Kim J, Pareschi L, Poyato D, and Soler J (2019) Vehicular traffic, crowds, and swarms: From kinetic theory and multiscale methods to applications and research perspectives. Math Models Methods Appl Sci 29(10):1901–2005
2. An J, Ryzhik L (2020) Global well-posedness for the Euler alignment system with mildly singular interactions. Nonlinearity 33(9):4670–4698
3. Aoki I (1982) A simulation study on the schooling mechanism in fish. Bull. Jpn Soc Sci Fish 48(8):1081–1088
4. Arnaiz V, Castro Á (2019) Singularity formation for the fractional Euler-Alignment system in 1D. https://arxiv.org/abs/1911.08974
5. Ballerini M, Cabibbo N, Candelier R, Cavagna A, Cisbani E, Giardina I, Lecomte V, Orlandi A, Parisi G, Procaccini A, Viale M, Zdravkovic V (2008) Interaction ruling animal collective behavior depends on topological rather than metric distance: evidence from a field study. Proc Natl Acad Sci U S A 105:1232–1237
6. Ballerini M, Cabibbo N, Candelier R, Cavagna A, Cisbani E, Giardina I, Orlandi A, Parisi G, Procaccini A, Viale M, Zdravkovic V (2008) Empirical investigation of starling flocks: a benchmark study in collective animal behaviour. Animal Behav 76(1):201–215
7. Blanchet A, Degond P (2016) Topological interactions in a Boltzmann-type framework. J Stat Phys 163:41–60
8. Blanchet A, Degond P (2017) Kinetic models for topological nearest-neighbor interactions. J Stat Phys 169(5):929–950
9. Bongini M, Fornasier M, Kalise D (2015) (Un)conditional consensus emergence under perturbed and decentralized feedback controls. Discrete Contin Dynam Syst A 35:4071
10. Caffarelli L, Vasseur A (2010) Drift diffusion equations with fractional diffusion and the quasi-geostrophic equation. Ann Math (2) 171(3):1903–1930
11. Caffarelli L, Vazquez LJ (2011) Nonlinear porous medium flow with fractional potential pressure. Arch Ration Mech Anal 202(2):537–565
12. Caponigro M, Fornasier M, Piccoli B, Trélat E (2015) Sparse stabilization and control of alignment models. Math Models Methods Appl Sci 25(3):521–564
13. Carrillo JA, Fornasier M, Rosado J, Toscani G (2010) Asymptotic flocking dynamics for the kinetic Cucker-Smale model. SIAM J Math Anal 42(1):218–236
14. Carrillo JA, Fornasier M, Toscani G, Vecil F (2010) Particle, kinetic, and hydrodynamic models of swarming. In: Mathematical modeling of collective behavior in socio-economic and life sciences, Model. Simul. Sci. Eng. Technol. Birkhäuser Boston, Boston, MA, pp 297–336

© Springer Nature Switzerland AG 2021
R. Shvydkoy, *Dynamics and Analysis of Alignment Models of Collective Behavior*,
Nečas Center Series, https://doi.org/10.1007/978-3-030-68147-0

15. Carrillo JA, Choi Y-P, Tadmor E, Tan C (2016) Critical thresholds in 1D Euler equations with non-local forces. Math Models Methods Appl Sci 26(1):185–206
16. Carrillo JA, Choi Y-P, Mucha PB, Peszek J (2017) Sharp conditions to avoid collisions in singular Cucker-Smale interactions. Nonlinear Anal Real World Appl 37:317–328
17. Carrillo JA, Choi Y-P, Perez SP (2017) A review on attractive-repulsive hydrodynamics for consensus in collective behavior. In: Active particles. Vol. 1. Advances in theory, models, and applications. Model. Simul. Sci. Eng. Technol. Birkhäuser/Springer, Cham, pp 259–298
18. Carrillo JA, Choi Y-P, Totzeck C, Tse O (2018) An analytical framework for consensus-based global optimization method. Math Models Methods Appl Sci 28(6):1037–1066
19. Castro A, Córdoba D (2008) Global existence, singularities and ill-posedness for a nonlocal flux. Adv Math 219(6):1916–1936
20. Cavagna A, Giardina I, Orlandi A, Parisi G, Procaccini A (2008) The starflag handbook on collective animal behaviour. 2: Three-dimensional analysis. Animal Behav 76:237–248
21. Cavagna A, Giardina I, Orlandi A, Parisi G, Procaccini A, Viale M, Zdravkovic V (2008) The starflag handbook on collective animal behaviour. 1: Empirical methods. Animal Behav 76:217–236
22. Cavagna A, Cimarelli A, Giardina I, Parisi G, Santagati R, Stefanini F, Tavarone R (2010) From empirical data to inter-individual interactions: unveiling the rules of collective animal behavior. Math Models Methods Appl Sci 20(suppl. 1):1491–1510
23. Cavagna A, Cimarelli A, Giardina I, Parisi G, Santagati R, Stefanini F, Viale M (2010) Scale-free correlations in starling flocks. Proc Natl Acad Sci U S A 107:11865–11870
24. Chae D, Córdoba A, Córdoba D, Fontelos MA (2005) Finite time singularities in a 1D model of the quasi-geostrophic equation. Adv Math 194(1):203–223
25. Choi Y-P (2019) The global Cauchy problem for compressible Euler equations with a nonlocal dissipation. Math Models Methods Appl Sci 29(1):185–207
26. Choi Y-P, Kalise D, Peszek J, Peters AA (2019) A collisionless singular Cucker-Smale model with decentralized formation control. SIAM J Appl Dyn Syst 18(4):1954–1981
27. Chuang Y-L, D'Orsogna MR, Marthaler D, Bertozzi AL, Chayes LS (2007) State transitions and the continuum limit for a 2D interacting, self-propelled particle system. Phys D 232(1):33–47
28. Constantin P, Vicol V (2012) Nonlinear maximum principles for dissipative linear nonlocal operators and applications. Geom Funct Anal 22(5):1289–1321
29. Constantin P, Drivas TD, Shvydkoy R (2020) Entropy hierarchies for equations of compressible fluids and self-organized dynamics. SIAM J Math Anal 52(3):3073–3092
30. Couzin I, Krause J, James R, Ruxton GD, Franks NR (2002) Collective memory and spatial sorting in animal groups. J Theor Biol 218(1):1–11
31. Cucker F, Dong J-G (2010) Avoiding collisions in flocks. IEEE Trans Autom Control 55(5):1238–1243
32. Cucker F, Dong J-G (2011) A general collision-avoiding flocking framework. IEEE Trans Autom Control 56(5):1124–1129
33. Cucker F, Smale S (2007) Emergent behavior in flocks. IEEE Trans Autom Control 52(5):852–862
34. Cucker F, Smale S (2007) On the mathematics of emergence. Jpn J Math 2(1):197–227
35. Danchin R, Mucha PB, Peszek J, Wróblewski B (2019) Regular solutions to the fractional Euler alignment system in the Besov spaces framework. Math Models Methods Appl Sci 29(1):89–119
36. Daneri S, Székelyhidi L (2017) Non-uniqueness and h-principle for Hölder-continuous weak solutions of the Euler equations. Arch Ration Mech Anal 224(2):471–514
37. DeGroot MH (1974) Reaching a consensus. J Am Stat Assoc 69:118–121
38. Dietert H, Shvydkoy R (2019) On Cucker-Smale dynamical systems with degenerate communication, to appear in Analysis and Applications
39. Do T, Kiselev A, Ryzhik L, Tan C (2018) Global regularity for the fractional Euler alignment system. Arch Ration Mech Anal 228(1):1–37

40. Favre A (1983) Turbulence: Space-time statistical properties and behavior in supersonic flows. Physi Fluids 26(10):2851–2863
41. Figalli A, Kang M-J (2019) A rigorous derivation from the kinetic Cucker-Smale model to the pressureless Euler system with nonlocal alignment. Anal PDE 12(3):843–866
42. Grassin M (1999) Existence of global smooth solutions to Euler equations for an isentropic perfect gas. In: Hyperbolic problems: theory, numerics, applications, Vol. I (Zürich, 1998), volume 129 of Internat. Ser. Numer. Math. Birkhäuser, Basel, pp 395–400
43. Ha S-Y, Liu J-G (2009) A simple proof of the Cucker-Smale flocking dynamics and mean-field limit. Commun Math Sci 7(2):297–325
44. Ha S-Y, Ruggeri T (2017) Emergent dynamics of a thermodynamically consistent particle model. Arch Ration Mech Anal 223(3):1397–1425
45. Ha S-Y, Tadmor E (2008) From particle to kinetic and hydrodynamic descriptions of flocking. Kinet Relat Models 1(3):415–435
46. Ha S-Y, Ha T, Kim J-H (2010) Asymptotic dynamics for the Cucker-Smale-type model with the Rayleigh friction. J Phys A 43(31):315201, 19
47. Ha S-Y, Kang M-J, Kwon B (2014) A hydrodynamic model for the interaction of Cucker-Smale particles and incompressible fluid. Math Models Methods Appl Sci 24(11):2311–2359
48. Ha S-Y, Kang M-J, Kwon B (2015) Emergent dynamics for the hydrodynamic Cucker-Smale system in a moving domain. SIAM J Math Anal 47(5):3813–3831
49. Ha S-Y, Kim J, Zhang X (2018) Uniform stability of the Cucker-Smale model and its application to the mean-field limit. Kinet Relat Models 11(5):1157–1181
50. Ha S-Y, Kim J, Park J, Zhang X (2019) Complete cluster predictability of the Cucker-Smale flocking model on the real line. Arch Ration Mech Anal 231(1):319–365
51. Haskovec J (2013) Flocking dynamics and mean-field limit in the Cucker-Smale-type model with topological interactions. Phys D 261:42–51
52. He S, Tadmor E (2017) Global regularity of two-dimensional flocking hydrodynamics. C R Math Acad Sci Paris 355(7):795–805
53. He S, Tadmor E (2020) A game of alignment: collective behavior of multi-species, to appear in Annales de l'Institut Henri Poincaré C, Analyse non linéaire
54. Huth A, Wissel C (1992) The simulation of the movement of fish schools. J Theor Biol 156(3):365–385
55. Isett P (2018) A proof of Onsager's conjecture. Ann Math (2) 188(3):871–963
56. Jadbabaie A, Lin J, Morse AS (2003) Coordination of groups of mobile autonomous agents using nearest neighbor rules. IEEE Trans Autom Control 48:988–1001
57. Kang M-J, Vasseur A (2015) Asymptotic analysis of Vlasov-type equations under strong local alignment regime. Math Models Methods Appl Sci 25(11):2153–2173
58. Karper TK, Mellet A, Trivisa K (2013) Existence of weak solutions to kinetic flocking models. SIAM J Math Anal 45:215–243
59. Karper TK, Mellet A, Trivisa K (2014) On strong local alignment in the kinetic Cucker-Smale model. In: Hyperbolic conservation laws and related analysis with applications, volume 49 of Springer Proc. Math. Stat.. Springer, Heidelberg , pp 227–242
60. Karper TK, Mellet A, Trivisa K (2015) Hydrodynamic limit of the kinetic Cucker-Smale flocking model. Math Models Methods Appl Sci 25(1):131–163
61. Kim J, Peszek J Cucker-smale model with a bonding force and a singular interaction kernel apparently the paper has not yet been published. So, just add (preprint) https://arxiv.org/pdf/1805.01994.pdf (2018)
62. Kiselev A, Tan C (2018) Global regularity for 1D Eulerian dynamics with singular interaction forces. SIAM J Math Anal 50(6):6208–6229
63. Kiselev A, Nazarov F, Volberg A (2007) Global well-posedness for the critical 2D dissipative quasi-geostrophic equation. Invent Math 167(3):445–453
64. Kuramoto Y (1975) Self-entrainment of a population of coupled nonlinear oscillators. In: Araki H (ed) International Symposium on Mathematical Problems in Theoretical Physics, Lecture Notes in Physics, vol 30, pp 420–422

65. Lear D, Shvydkoy R (2019) Existence and stability of unidirectional flocks in hydrodynamic Euler Alignment systems, to appear in Analysis & PDE. https://arxiv.org/abs/1911.10661

66. Lear D, Shvydkoy R (2020) Existence of unidirectional flocks in hydrodynamic Euler Alignment systems II: singular model, to appear in Communications in Mathematical Sciences

67. Lear D, Reynolds D, Shvydkoy R (2020) Grassmannian reduction of Cucker-Smale systems with application to opinion dynamics. Preprint

68. Leslie TM (2019) Weak and strong solutions to the forced fractional Euler alignment system. Nonlinearity 32(1):46–87

69. Leslie TM (2020) On the Lagrangian trajectories for the one-dimensional Euler alignment model without vacuum velocity. C R Math Acad Sci Paris 358(4):421–433

70. Leslie TM, Shvydkoy R (2019) On the structure of limiting flocks in hydrodynamic Euler Alignment models. Math Models Methods Appl Sci 29(13):2419–2431

71. Mao Z, Li Z, Em Karniadakis G (2019) Nonlocal flocking dynamics: learning the fractional order of PDEs from particle simulations. Commun Appl Math Comput 1(4):597–619

72. Markou I (2018) Collision-avoiding in the singular Cucker-Smale model with nonlinear velocity couplings. Discrete Contin Dyn Syst 38(10):5245–5260

73. Minakowski P, Mucha PB, Peszek J, Zatorska E (2019) Singular Cucker-Smale dynamics. In Active particles, Vol. 2, Model. Simul. Sci. Eng. Technol. Birkhäuser/Springer, Cham, pp 201–243

74. Morales J, Peszek J, Tadmor E (2019) Flocking with short-range interactions. J Stat Phys 176(2):382–397

75. Motsch S, Tadmor E (2011) A new model for self-organized dynamics and its flocking behavior. J Stat Phys 144(5):923–947

76. Motsch S, Tadmor E (2014) Heterophilious dynamics enhances consensus. SIAM Rev 56(4):577–621

77. Mucha PB, Peszek J (2018) The Cucker-Smale equation: singular communication weight, measure-valued solutions and weak-atomic uniqueness. Arch Ration Mech Anal 227(1):273–308

78. Niizato T, Murakami H, Gunji Y-P (2014) Emergence of the scale-invariant proportion in a flock from the metric-topological interaction. Biosystems 119:62–68

79. Olfati-Saber R (2006) Flocking for multi-agent dynamic systems: algorithms and theory. IEEE Trans Autom Control 51(3):401–420

80. Park J, Kim HJ, Ha S-Y (2010) Cucker-Smale flocking with inter-particle bonding forces. IEEE Trans Autom Control 55(11):2617–2623

81. Perea L, Elosegui P, Gomez G (2009) Extension of the Cucker-Smale control law to space flight formations. J Guid Control Dynam 32:526–536

82. Peszek J (2014) Existence of piecewise weak solutions of a discrete Cucker-Smale's flocking model with a singular communication weight. J Differential Equations 257(8):2900–2925

83. Peszek J (2015) Discrete Cucker-Smale flocking model with a weakly singular weight. SIAM J Math Anal 47(5):3671–3686

84. Piccoli B, Rossi F, Trélat E (2015) Control to flocking of the kinetic Cucker-Smale model. SIAM J Math Anal 47(6):4685–4719

85. Poupaud F (1999) Global smooth solutions of some quasi-linear hyperbolic systems with large data. Ann Fac Sci Toulouse Math (6) 8(4):649–659

86. Poyato D, Soler J (2017) Euler-type equations and commutators in singular and hyperbolic limits of kinetic Cucker-Smale models. Math Models Methods Appl Sci 27(6):1089–1152

87. Reynolds CW (1987) Flocks, herds and schools: A distributed behavioral model. ACM SIGGRAPH Comput Graph 21:25–34

88. Reynolds DN, Shvydkoy R (2020) Local well-posedness of the topological Euler alignment models of collective behavior. Nonlinearity 33(10):5176–5215

89. Shang Y, Bouffanais R (2014) Consensus reaching in swarms ruled by a hybrid metric-topological distance. Eur Phys J B 87(12):294

90. Shu R, Tadmor E (2021)   Anticipation breeds alignment. Arch. Rational Mech. Anal. 240:203–241
91. Shu R, Tadmor E (2020) Flocking hydrodynamics with external potentials. Arch Ration Mech Anal 238(1):347–381
92. Shvydkoy R (2019)   Global existence and stability of nearly aligned flocks.   J Dynam Differential Equations 31(4):2165–2175
93. Shvydkoy R, Tadmor E (2017) Eulerian dynamics with a commutator forcing. Trans Math Appl 1(1):tnx001
94. Shvydkoy R, Tadmor E (2017) Eulerian dynamics with a commutator forcing II: Flocking. Discrete Contin Dyn Syst 37(11):5503–5520
95. Shvydkoy R, Tadmor E (2018) Eulerian dynamics with a commutator forcing III. Fractional diffusion of order $0 < \alpha < 1$. Phys D 376/377:131–137
96. Shvydkoy R, Tadmor E (2020) Multi-flocks: emergent dynamics in systems with multi-scale collective behavior, to appear in Multiscale Modeling and Simulation
97. Shvydkoy R, Tadmor E (2020)   Topologically-based fractional diffusion and emergent dynamics with short-range interactions. SIAM J. Math. Anal. 52(6):5792–5839
98. Silvestre L (2012) Hölder estimates for advection fractional-diffusion equations. Ann Sc Norm Super Pisa Cl Sci (5) 11(4):843–855
99. Tadmor E, Tan C (2014)   Critical thresholds in flocking hydrodynamics with non-local alignment. Philos Trans R Soc Lond Ser A Math Phys Eng Sci 372(2028):20130401, 22
100. Tan C (2019)   Singularity formation for a fluid mechanics model with nonlocal velocity. Commun Math Sci 17(7):1779–1794
101. Triebel H (1995) Interpolation theory, function spaces, differential operators, 2nd edn. Johann Ambrosius Barth, Heidelberg
102. Vicsek T, Zefeiris A (2012) Collective motion. Phys Reprints 517:71–140
103. Villani C (2003)   Topics in optimal transportation, volume 58 of Graduate Studies in Mathematics. American Mathematical Society, Providence, RI

Index

© Springer Nature Switzerland AG 2021
R. Shvydkoy, *Dynamics and Analysis of Alignment Models of Collective Behavior*,
Nečas Center Series, https://doi.org/10.1007/978-3-030-68147-0

Printed in the United States
by Baker & Taylor Publisher Services